300 YEARS OF INDUSTRIAL DESIGN

Published in the United States in 2000 by
Watson-Guptill Publications
a division of BPI Communications, Inc.
1515 Broadway
New York, NY 10036

First published in Great Britain in 2000 by
The Herbert Press, an imprint of
A & C Black (Publishers) Limited
35 Bedford Row, London WC1R 4JH

Library of Congress Catalog Card Number: 00-100119

ISBN 0-8230-5368-7

Cover design by Marie Mundaca
Front cover illustration: Edison's light bulb
Back cover illustrations: Model T Fords in production; Portsmouth blockmaking machinery; night lamp (taken from Wedgwoods' 1880 shape catalog)

Fold-out-orientation chart by SIGNAL GRAFIK Aps, Randers, Denmark

Typeface: Times New Roman; 10.5pt
Paper: Fineblade extra, 130gsm
Cover: one sided artboard, 220gsm

Printed and bound in Great Britain by Hillman Printers Ltd, Frome, Somerset

First printing, 2000
1 2 3 4 5 6 7 8 9 / 07 06 05 04 03 02 01 00

300 YEARS OF INDUSTRIAL DESIGN

Function · Form · Technique 1700–2000

Adrian Heath
Ditte Heath
Aage Lund Jensen

consultant for ceramics and glass
Snorre Læssøe Stephensen

WATSON-GUPTILL PUBLICATIONS/NEW YORK

Dedicated to
Aarhus School of Architecture, Denmark,
and to its students, past, present and future.

Acknowledgements

The research, writing, drawing and photography for this book have been made
possible by funds and support from the following Danish institutions:

Aarhus School of Architecture
Denmark's National Bank's Anniversary Foundation of 1968
Margot and Thorvald Dreyer's Trust
Consul George Jorck and Mrs Emma Jorck's Trust
Ellen and Knud Dalhoff Larsen's Trust

The book is the work of many hands: helping hands at the factories, work-
shops, industrial sites, design schools, libraries and museums of the countries
we have visited. To all these, and to friends and colleagues, we are indebted.
We wish to thank all those past students of the Aarhus school of Architecture
who have taken part in the project and done measured drawings, and especially
Claus Bech-Danielsen, Michael Heath, Charlotte Lemme, Akis Nakagawa,
Ulrik Sterner Nordam, Annette and John Wessman and Nis Øllgaard. Thanks to
Martin Bohøj for his great contribution, and for sticking with us all the way.

We are indebted to John and June Broome for help with an important item of
research and correction of texts, and to David Whiting for trying to improve
our English and for his encouragement. Thanks to those who have lent artefacts
for us to draw and photograph; to Chris Jones, of Posford Duvivier, for help
with the Palm House; to Peter Hollins for introducing us to the Block Mills at
Portsmouth Naval Dockyard; to Ole Bang for assisting us with the work
of Michael Thonet; and to Ione Heath and John Morton for their
active interest in the project seeing the light of day.

CONTENTS

From the exhibition 'Anonymous Design' at Louisiana Museum of Art, Denmark, in 1974.
Photographed by Aage-Lund Jensen, the photographer for the book, unless otherwise stated.

INTRODUCTION

*The artist must never forget that he is a technician,
the technician never that he is an artist.*

An old aphorism

Observation is a necessary part of creation

This is as much a designer's notebook as it is a history book. The authors are not historians but designers and architects, and much of the research on which the material is based was conceived for teaching – and self teaching. We have laid out an industrial history of design by looking closely at some of the best utility artefacts used in daily life and some of the greater inventions, buildings and machines which mark out the course of industrialisation. We regard design as a kind of craft and see the manufacturing complex as an enormous and fascinating design development which tells us about the way human needs and activities, form and use, technique and materials each take equal parts in what is known as industrial design. We wanted to stage a meeting between the craft of designing and the craft of manufacturing. This is a book which deals directly with the job of designing; it is not about design policy, or design as a sales factor, or even about designers. It is as much as anything about how to observe the good things around us and to do it with a purpose.

Making the book (we feel it to be as much 'made' as written) has taught us a tremendous amount not only about history and its erratic course, but that industrialisation has both given and taken away – it has made us collectively capable but personally incapable – and left us with an accumulation of environmental damage which affects every single part of life. Raking through and highlighting some of the pieces of industrial design history has shown the good and exposed the bad.

The research

Our research started in spring 1974 when we left Denmark with a car load of photographic, measuring and drawing equipment bound for Newcastle and the oldest industrial regions of England: Durham, Sheffield, Ironbridge, Birmingham and London. Our mission was to record some of the best early industrial products. Behind us we had left several months of preparation in the form of correspondence with museums and factories as well as designing and making photographic and graphic recording equipment. Similar research trips have since been made, with and without students, to Stoke-on-Trent, High Wycombe, Portsmouth, Munich, Paris, Amsterdam, Oslo, Sweden, Helsinki and venues in Denmark. This work has resulted in a large collection of drawings, photographs, slides and artefacts from the whole industrial period, much of which has been used in the book.

INTRODUCTION

The technological phases

This has always been a very practical project. It has been important for us to experience the artefacts, whether buildings, machines or products, by direct contact and to record them often in full size. Only a small minority of the items in the book are untouched or untested. In our experience, few books have seriously tried to make the facts, skills and feeling of design available to the reader. They have mostly relied upon text and photographs alone, which are not of much use to students of the design process. But we have to say that it was largely thanks to a book, Lewis Mumford's *Technics and Civilization*, that we started this undertaking. With some inspiration from Patrick Geddes, Mumford divided the history of industrial civilisation into three overlapping and interconnected technological phases, each with its specific regions, resources, raw materials, forms of energy, forms of production and workers. He called the phases eotechnic: the phase of water and wind power and natural materials; palæotechnic: the phase of steam, coal, iron and the beginnings of mass production and neotechnic: the phase of electricity, the internal combustion engine, alloys and plastics. Each period refined the achievements of the preceding period. The book was written in 1935 so the phase of computerised globalisation, whatever he might have called it, is not covered. To the industrial designer this is a very telling way of looking at the history of design. It forces us not only to evaluate the form of the finished product, but also to question what it is made of, and how and why, and to relate the product to other products and other industries. One places the product in its technological context and not merely according to its appearance, style and date. It is not strictly speaking a scientific classification, but once one understands it, design-historical evaluations can be made at a glance. For example, it is Mumford's classification that justifies our placing of the trug in the early eighteenth century. Although our example was made in the 1970s, it is still made by hand with natural materials much as it would have been in the eighteenth century. It is as useful and beautiful now as it was then.

Comprehensive as well as detailed

From the outset we were determined that when on location we would not only interest ourselves in the products, but also in the technology and architecture of which they are a part; that is, their conditions of use, production processes and machine tools, the prime movers involved, transport and the factories and workers' housing where these were of interest. In this respect we received much help from the industrial archaeologists and the 'live' museums, especially in Great Britain. This approach has, we hope, given the book both breadth and depth. It is responsible for product choice and for much of the material in Part I, Technological Background, which we see as data and a context for the products of metal, wood, ceramics and glass that follow.

The construction of the book

The idea and construction are very simple, though a little unusual. A glance at the orientation chart at the end of the book clarifies the construction and to some extent the idea of the work: the chronology moves from left to right horizontally in decades; materials are placed vertically. Each double page in each part (with the exception of Part I) deals with one product or product type which on the chart is shown as a drawing of the product. The eighteenth century is divided not into decades but, because of the difficulty in finding and recording products of this age with the same degree of detail as the later ones, into three periods only.

This we regret because although there was less industrial activity at that time, there was more than one would think, and of a particularly fine quality. Part I is very different from the other sections, and is essentially a service section. It lists the historical events of the decade (or period) which, directly or indirectly, have influenced the products in the other Parts and gives them their industrial context. The listed events or innovations which have been of particular relevance to society and industrial design are taken up for further discussion and illustration.

Design archaeology: the importance of the measured drawing

There was another coercive factor which put us on the road. In the Danish design tradition there is a respect, unknown in most countries, for the observation and recording of worthwhile artefacts by measured drawing. It is a design/educational discipline. Doing a measured drawing of a chair, for example, like those in Part III of the book makes one observe every detail; it puts one through many of the mental and physical activities which are necessary to design a chair; it explains to the student the full meaning of the designer's principal working media, the plan, the section and the elevation; it teaches drawing techniques and control. It is exciting to handle and record good, sometimes famous, artefacts. The practice and use of the technique of measured drawing is one of the most important features of the book and is discussed further in this and the other introductions.

Photography

For our purpose the drawings need the support of the photographs. While the measured drawings show the exact and objective truth about an artefact, the photo expresses the essential 'feeling' of it. The object is brought to life by the light in which it stands. We see its form and texture and the multitude of nuances which combine to convey its being. In the measured drawing is the secret to understanding just why an artefact is 'right' – an explanation of some of the subtleties of the photograph. Many are field photographs taken in the difficult conditions of the location in which the objects were found. However, our aim has been to make the artefacts as 'touchable' as possible by good photography – most of which is original. The photographs are black and white, rather than colour, because we have found that it allows better concentration on form and surface.

Using the book

The best way to introduce oneself to the book is to fold out the orientation chart inside the back cover. Let it lie to the right of the book while turning the pages. The artefacts illustrated and discussed in Part I (Background Technology) will be found in the top horizontal strip of illustrations on the chart. The artefacts in Part II (Metal Products) in the next strip, and so on through the Parts dealing with wood (III), ceramics (IV) and glass (V). With the help of the orientation chart the reader can also get a picture of the relative stages of design in the different materials in the same decade. Occasionally, artefacts have a direct relationship with one another, as with the Blockmaking Machinery, 1800–10, in Part I and their product, the pulley block (Part III). These are printed in the same vertical column on the orientation chart.

Most of the product pages comprise measured drawing, photograph and text. The reader can let the drawing and the photograph support each other so that, with the

INTRODUCTION

text, we hope that it is possible to understand everything one needs to know about the product. Although it is impossible to provide a full substitute for the real thing, the measured drawings do in fact reveal information which is hard to see on the object itself.

Our research has made considerable use of old catalogues and of encyclopaedias. The etchings with which they are illustrated are often beautiful pieces of craftsmanship as well as being informative about their subject, and about the technological and commercial status quo. We have used these occasionally when we have been unable to find the actual artefact, or if that artefact has ceased to exist. But, they have to be taken with a pinch of salt, especially if they were made for commercial reasons. They are generally not true to scale and sometimes indulge in a certain amount of artistic licence.

Product choice and chronology

This book is not an attempt to systematise the history of industrial design. That would be unmanageable and not very useful. What we have done is to break the subject down into types of manufacturing materials and superimpose it onto the decades of time. The choice of individual products has been governed by a particular attitude to design and the luck involved in searching and finding. The majority of the products have been designed by people who, usually craftsmen themselves, have shaped their objects with sensitivity, perception and consideration for the user. This integral form sense has dictated our choice. The products have something important to tell us about the relationship between function, form and technique.

The placing of products is chronological, but the chronology of objects produced in large quantities which have been made over a longer period, or dropped and later resumed, naturally has to be flexible. Sometimes they are placed in the decade of their invention, sometimes when they were at the height of their popularity. The first is of historical interest; the second tells us of the success of the product. More important, for us, than exact chronological placement, has been the extent to which we can learn something about design from them.

When we put these products into categories according to the materials from which they were made – with all their variables of country and origin – and then onto a grid of decades with one product of each material per decade, we get a number of interesting results. Product differences, or lack of them, become evident by comparison with the regular beat of time. To give some examples: we suddenly realised that homogeneous things like glass products have barely changed in 300 years. Who would have thought that plant cultivation was assisted by the glass cloche in the early 1700s? And that eotechnic Danish chairs appear as chronological neighbours to a palaeotechnic English railway wheel? And that Diderot's Encyclopaedia of the 18th century deals in detail with screw thread making tools, and the thread making slide rest, when in fact one hundred years were to pass before somebody thought of standardising the screw thread? There are many such intriguing examples of the incongruities of the historical development of industrial design.

What is a product?

The word 'product' as applied to utensils is problematic because it implies that they are merely objects which have been produced, not things to be used. The Scandinavian word *brugsting* is much more telling. It means simply 'use things'. The English language has nothing so direct. We get diverted into using phrases such as 'applied art' or words such as 'artefact' which miss the point. 'Utensil' is a good word which is self explanatory but we associate it with pots and pans. So, to avoid confusion, we will usually continue to use the industrially-orientated word 'product', and fall back on 'artefact' as a general term for anything manmade.

Some characteristics and tendencies of industry
Each of the descriptions of 160 artefacts in Parts I to V, and the 890 entries in Part I outlining background technology, carry not only design and technical information but they also tell us something about the characteristics of the industrial complex generally, and its relationship to design and the designer, ecology, materials, design education, and design history. The following notes on these aspects are in no way exhaustive, but are simply general observations made while working with the subject material.

From our 1990s vantage point we can discern four overlapping periods in the way things have been designed, made, distributed and sold: (1) the period of large quantities of handmade items produced in the workshop and home of the 18th and early 19th centuries with largely direct sale; (2) the period of early industrialism with machine-aided handwork in larger factories (and in homes) up to the beginning of the 20th century with speculative, more organised sale; (3) the period of a much higher degree of mechanisation in large units up to the 1940s and 1950s; and (4) the period of a high degree of automation and computerisation, and finance guided globalisation, with a distribution and marketing industry which must be nearly as big as the manufacturing industry itself.

The tendency throughout the last 300 years has been towards ever greater size, and use of energy, so that in spite of cleaner technology and better working conditions the relatively slight pollution of long ago has now escalated and threatens the global environment. This situation is beginning to be acted upon politically and through new movements of popular consumer action.

The causes of this escalation are fairly clear. They are not explained by economics alone, but by the much more fundamental nature of production. As soon as you start making things of which many are needed, you very soon learn that larger quantities in production – size of batches, size of deliveries leading to size of machines and buildings – is in itself a production advantage. The increase in size is a natural law of production whether it be within product manufacture, the winning of raw materials, agriculture or the food industry. It started in the workshop and because the development of the machine was always there to help, has grown to the gigantic production units of today. It is commonly known as 'progress' although it often has little to do with quality.

The whole question of specialisation is important – not least in the context of design. Specialisation of factory operations led to division of labour which led to specialisation of factories. From the point of view of production this was, and is, a great advantage. It quite simply increases output. Again, it is almost entirely a rational tendency which originated in the workshop and it seemed, when first introduced, like a law of production. We are now learning that from the point of view of human well-being and development, it has very negative effects on both mind and body.

INTRODUCTION

Factory specialisation, where different companies manufacture the various components of a product, which are then assembled by the marketing company, is today the commonest type of production for composite products. All this is a matter of economics alone. To those who believe that industrial design and making are fundamental human activities which should be comprehensible to those directly involved, and as far as possible to people at large, the latest signs of a move away from specialisation, and towards a greater degree of participation, are encouraging.

Another basic tendency of industry has been that of a fairly steady, and escalating, increase in the achievement of certainty in manufacture – from being one of continual checking of processes and machine operations towards one of automation. The artefacts, here shown in their chronological order, demonstrate fairly clearly the gradual movement away from designing and making by judgement and personal skill towards mechanised precision and repetition. The first gave variation in, and between, products and buildings. The second gave us the environment of uniformity which lacks variation in surface textures and the contrasts of light and shade. The manufactured environment has tended to flatten and deny the senses, and has gradually reduced personal skill.

Industrialisation has often robbed materials of their beauty, not because it is necessarily its nature to do so, but because consideration of truth to materials gets ignored in the eagerness to increase sales. There is very often a design decision to be taken here where there is no designer to take it. Our synthetic world has lost respect for the genuine article and is very good at substituting this by unrelated, inappropriate patterning or some other camouflage. One only has to mention 'wooden', or 'marble' plastic laminate, or artificial leather. There are hundreds of examples: the components of the Windsor chair were made of various woods which were chosen according to their natural properties in manufacture and use – elm for the seat, beech for the legs, ash for the back bow and so on. Instead of allowing these woods to show their natural qualities of grain and colour the whole chair was stained dark brown. The various visual characteristics of the woods were lost.

Working concurrently with industrial growth, and the disguising of products to look like anything but what they actually are, has been the support from advertising. Here one of the favourite expressions of mass production used to be, and surprisingly still is, 'homemade'! It is falseness in much of industrial production which, by comparison, makes the functional tradition so important – and so attractive.

Which brings us to what we now call the production engineer, for he was responsible for much of the industrial functional tradition. All through the development of industry the production engineer has been, and is, a key person. One seldom hears anything about him but much of the best industrial design thinking has been devoted to the machinery of production itself. It is the production engineer (often without the fine title) who is largely responsible for the development we trace here. He was understandably given a great deal of power by the factory owner. It is after all production which is the most important part of the whole process. The early engineers were remarkably good designers. They worked within the traditions of the Renaissance, but in a period and atmosphere of innovation, the dawn of a quite new technology, added to which they were themselves skilled craftsmen. It must have been a very exciting time. There are a number of examples of their work shown and discussed here, the hand press (pp 78-79), the

Portsmouth blockmaking machinery (pp 34-35), the work of Christopher Polhem (pp 28-29), and the inventor and engineer, Jaques de Vaucanson, are just some of the most remarkable.

So the tradition of craftsmen designers was not only to be found amongst potters, glassmakers and chairmakers but, at the core of the Industrial Revolution, among the engineers.

An industrial development which started in the late nineteenth century and which has escalated ever since is what one might call design-to-product automation. It developed from automation for the production of very simple things which were required in very large quantities such as screws, nuts and bolts, nails and bottles. Increased demand (often created merely by increased marketing) called for machines and machine sets to make one particular product so that the design of the product and of the machines to make it became a single design and manufacturing project. There are many examples of this in the things around us, especially in packaging, for example the folded cardboard milk carton. Product machine manufacturing companies now distribute their machines all over the world and the product, patented along with the machine, is as they have decided it should be: the same everywhere. These historically newer methods make manufacturing descriptions almost impossible. All the details of making remain the secret of the machine. In our manufacturing descriptions we have therefore tried to keep to the principles involved and the basic operations which in fact take place. This sort of description usually makes processes sound much slower than they actually are.

With size and quantity has come power and momentum. Some of the major contradictions of industrial civilisation are well illustrated by one man and his factories. Henry Ford's social, design, and manufacturing methods, demanding as they did production and distribution planning on a hitherto unknown scale, stand in sharp contrast to the chaotic condition of the road system on to which his 15,000,000 'T' cars were let loose. The whole road transport project of modern times has always had the character of the unreal, the impossible. The railway had to be planned, the motor car never was.

Nearer to our particular subject we can see the motor car as the prime creator of the mass production, sales and commercial attitudes which have been with us ever since its arrival – especially that of the cultivation of obsolescence as a means of artificially increasing production and sales.

Since research started for this book in the 1970s, industry, commerce, industrial design, and communications have changed radically, as can be seen from the last pages of Part I. What does not feature there is the globalisation which has taken place in these 25 years. It is now quite common for a product to be designed in one country, manufactured in a second (usually in east Asia), from materials from a third, assembled in a fourth and marketed from a fifth. And this arrangement is competitive, costing less than if the product had been made entirely in the initiator's country. Monetary price is the sole criterion. All the cultural, ecological and socially destructive elements in the operation – the real costs – remain unseen, unquestioned.

The environmental aesthetics of our industrial civilisation will not be discussed further here, except to mourn the loss of the only direct contribution that industrialisation seems to have made to the sensing of the natural elements: the evocative tone of the wind in the telephone wires. Now even they are disappearing, and being put underground!

The design-making synthesis

The intensely difficult task of the designer is to have a deep understanding of the industry for which he is working and, at the same time, perform the role of advocate for humanity and its environment. No small task!

Before discussing this relationship, it helps to have a clear picture as to what designers are, for there are almost as many different types of designer as there are industries. The five principal groups we are interested in here each have quite separate roots and they each work under different conditions. They use a different professional language, have different attitudes, different aims with their work and are largely unknown to each other. The work of all of them is represented in this book.

The craftsman designer is the most natural of them all, and has roots reaching right back into pre-industrial history. They are engaged with all the materials of this book but usually specialise in one group of related materials only. They have become fewer and fewer throughout our story, although there now seems to be the beginnings of a renaissance. They will always exist because their way of doing things is the most obvious. They design what they make, and these two processes often overlap. With regard to knowledge of materials, and especially his or her own, they will always have the edge on the industrial designer.

The industrial designer is very different from the craftsman, although in many respects it would be good if they had more in common. Depending on the type of company and product, the person who designs can be a works foreman, a mechanical engineer, a salesman, or the director of the firm. For some products this absence of design emphasis is obvious, but for most products that are operated, or used by people, it is unwise not to use a designer. Industrial designers are trained to express ideas, and construct, by drawing and model/prototype making; they have the knowledge and insight to design products so that they perform the practical, visual and tactile functions required by those who should use and maintain them; and they are concerned with the quality and economy of the manufacture of the product. This manifold process can only be carried out successfully if there exists close co-operation between the designer and the manufacturer. That is to say all the people named above. This often poses difficulties because of divergent interests. The manufacturer is responsible for a successful and economical production and distribution. The designer's responsibility is the functional and æsthetic content of the product. The optimal situation occurs when this difference of interests becomes a productive synthesis.

The architect designer. In some countries, Scandinavia for example, the English word 'designer' has only come into use comparatively recently. He or she was an 'architect' and the word 'architecture' is still used in relation to well-designed products. The schools of architecture all have design departments. Considering that a large proportion of product design work is related to building this is a very good background for a designer to have: a product not only has to be made and shaped well, it generally has to relate to people, the users, and become part of an environment. This should be considered as part of the design process.

We have already discussed the key position of the production engineer, and his part in the design process. But some of the best artefacts, especially in Part I, such as bridges, work buildings and prime movers, were designed principally by structural and mechanical engineers. Some early engineering design was remarkable. Good examples are the Iron Bridge (pp 32-33), the steam hammer (pp 40-41), and the steam pump at the Ryehope waterworks (pp 48-49).

A fifth designer group which is not so relevant to this book, but is none the less important, is the artist designer; the consultant painter or sculptor who was called in, mostly by manufacturers of domestic wares such as pottery and glass, to advise on the artistic content of a product and to design pattern, and sometimes shape it. Before the industrial designer came on the scene this was the only option, but since the 1920s the artist has been used less and less in industry. The Victorian period was the heyday of the art consultant. It was important to make products look more than they really were, and to conceal the fact that they had been mass produced. The results were usually false and dreadful – but occasionally magnificent. Historically, the phenomenon is interesting because the idea of applying art to a utensil – as opposed to the idea of making it beautiful in itself, through its own function and form – was the notion from which the industrial designer was born. This did much to confuse the industrial and commercial world as to the real meaning of industrial design, and is a British phenomenon. In Scandinavia, for example, industrial design education developed from the crafts and technical schools, and the schools of architecture, not from art schools.

With the complexity and scale of design work, collaboration has become increasingly important, not only between people who are directly concerned with the design and manufacture of a product but, in the early stages of design development, with quite other professions with special knowledge about the users and use of the project in question.

Throughout the last 130 years the act of designing and that of making have moved further and further away from each other. The designer knows less and less about production and the producer knows less and less about the real meaning of design. One of the objects of this book is to bring the designer and the producer a little closer again.

Design history and the functional tradition

The book breaks down design history into individual products, sorted according to the primary material of which they are made, and keeps the heavier background technology separate but connected by the progression of decades which are common to the entire contents. One of our reasons for doing this is to bring some sort of order into a highly diverse subject, and to nurture concentration. We have made a point of not expecting the reader to evaluate a water mill and a chair, or an aeroplane and a lamp on the same page. We find that this mixing of subjects is a problem for the designer/reader – and for design – in many design history books.

Somebody once said 'history is a sieve'. It was a Danish cellist who was discussing the way in which the best of musical instrument design rises to the top and remains there over hundreds of years. His short, slightly startling remark is applicable to all industrial design history and to a certain extent to this book. By also being a kind of sieve it sorts out some of the more design-relevant artefacts for further observation and investigation. When we started we had high hopes of selecting only history's key products: the turning points in the history of industrial design, but we soon discovered that the mesh in our sieve was too fine and we had to accommodate a whole range of examples of which only a few are key works. Examples of what can be described as 'key artefacts' taken from each part of the book include the Bramah Lock of the 1780s, Thonet's bent wood chairs of the 1850s, the traditional ewer shown in the 1890s, and the Palm House at Kew from the 1840s. They are all still in use and some are still in production. In Part I all the illustrated subjects can be described as key artefacts.

A glance at the fold-out orientation chart at the end of the book will show that we are not interested in the ornamental but rather in the mainstream of products which were made to be durable (physically and visually), inexpensive and useful. It is what we can call the functional tradition and as the phrase implies it has always been with us. However, phrases like this have a way of being used too loosely, so a simple definition might be helpful: buildings, machines and products which have been thoughtfully designed and made, often over a long period, to fulfil certain functions and uses, and so possess an intrinsic beauty of structure, craftsmanship and form. The phrase applies mostly to the architecture and design of work and daily life. One of the very few museums to specialise in the functional tradition is Die Neue Sammlung, in Munich.

The functional tradition plays an important part in design history because it lies at the core of the design discipline. This area of everyday, good design, although constituting only a small percentage of production, is that which affects most people. But quite another, much smaller, stream of activity, better known in design historical literature, had surprisingly little influence on it. This was the architect/ designer/ painter/ sculptor and craftsman-initiated movement which arose as a (more or less) organised reaction to the harsh realities of rapidly developing industrialisation towards the end of the nineteenth century. It quite naturally started in England, where industrialisation had been developing longest, with the group led by William Morris and Phillip Webb. In design histories the movement is often traced right through to the German Bauhaus via Wiener Werkstätten, Deutscher Werkbund, Deutscher Werkstätten, the Dutch De Stijl and a number of other initiatives. Seeing these movements in a continuous line, of some 60 years duration, is very telling because they all believed in the integration of designing and making on a high level and they were all critical of prevailing design and production. But whereas Morris's work remained a protest against mechanisation, some of the Austrian and German organisations actually thought and worked industrially and their members were, in fact, to influence industrial design. In these pages their work is represented by Peter Behrens (Deutscher Werkstätten, pp 104-105) and Wilhelm Wagenfeld (Bauhaus, pp 152-153 and pp 260-261). The historical curiosity of all this is that our interest in the functional tradition in design, based on industrial and technological development, actually 'bypasses' the protest based work of the 'Arts and Crafts' movements of Europe. This has to do with the growing separation of the industrial and commercial world, and the world of the designers and craftsmen. Their ideas were sound and their aims exemplary, but they were incompatible with the greater part of everyday industry with which we are concerned here. There has to be a prime interest on the part of both makers and marketers, in good design-and-manufacturing quality.

In 1924 – while the Bauhaus, under Walter Gropius, was exploring the new ideas of 'functionalism' – a little way north in Copenhagen, Kaare Klint was starting a new department of furniture and interior design in the School of Architecture at the Royal Academy. Here the students were also investigating functional design, but in quite a different way. They looked closely at the functional furniture and architecture of history by observation and recording, through accurate measured drawing, in full size, and analysed just why these designs were so comfortable, useful, strong and beautiful. The drawing of a Shaker rocking chair on p. 144 is an example of such a study by one of Klint's students. After they had made these analyses they set about interpreting the results by designing new furniture for modern living and machine production. The design of storage furniture began with a survey and measurement of the articles to be stored and studies of how best these articles could be arranged to be economical of space and accessible. All furniture design was accompanied by studies of the dimensions and movements of people and these studies helped to established the size and shape of rooms. This research and design activity was part of the foundation of what could be called the functional tradition in new Danish furniture, industrial design and architecture, which started in the late 1920s.

Almost half of the artefacts we have examined are from Britain. This is not only because industrialisation started here, but also because this country is a mine of buildings and products of the functional tradition. Although much has been destroyed to make way for new – and often inferior – development, and you have to search for it, there is much left which 'live' museums and an active industrial archaeological movement have saved, restored and made accessible. Ancient water-powered and steam-powered factories and services have been made to work and live again.

Each period of industrialisation has its own special atmosphere, from the primitive roughness of early water-powered factories (see Wortley Forge, pp 136-137) to the impressive sophistication of later steam power (see Ryehope Pumping Station, pp 48-49), and to the products of early electricity which seem to tell us both where we have just been and where we are going (see stable lamp, pp 200-201).

We have found that industrialisation has to some extent been the process of making specialised equipment for the laboratory and factory available to the general public. Many more products than one might think have originated in this way, and they have often been among the best, because their design is based on real needs and functions. This adaptation of the professional tool to the general market has occurred throughout the 300 years. Typical examples are the thermometer and the computer.

Design and ecology
It is only during the last 30 years that we have gradually become concerned about the effect of industrialisation on the environment. When research for this book started in the 1970s most of us did not question the fundamentals of the way technology was taking us. In fact, it had improved in so many ways – in its products, in its working conditions and in what it meant for people's daily lives. Now, at the end of the century, there is not only popular concern for the environment, but action is being taken by governments and environmental organisations worldwide as well as by a number of enlightened manufacturers.

Far from attempting to observe industrial ecology, our mission here has been, more simply, to record some of industry's everyday, useful products. However, this chronology also provides a foundation on which to judge the deterioration of the environment. Part I, Background Technology, gives a useful exposition of what has happened in history. Here we can clearly see the 'snowball' effect of industrial activity on the environment – a process which has been going on longer than is commonly thought.

Clearly designers should be key people in this situation. The question is whether the forces and ways of the free market will permit them to be so. Ecological thinking is extraordinarily difficult. It is mostly concerned with changing practice in our everyday ways of doing things – often quite simple actions which previously had happened naturally and unconsciously, but which now, due to our industrial welfare, have become strangely elusive. Much of it has to do with simply economising with materials and energy; something which we always used to do when we had less to do it with, and were poorer and less well equipped. One of the best ways to design ecologically is to make things so well that, instead of being replaced, they remain in use for many years and become part of our cultural heritage – like the products in this book. Good designing is, to a large extent, putting ecology into practice.

Each design field has its own set of ecological rules and each particular design task its own requirements, but the design/ecological realities of a product or building have to be discussed with the manufacturer or client in the early stages of planning. This concerns the critical analysis and choice of raw materials, including the amount of energy and transport involved; the economy and viability of production – again, including the amount of energy and transport involved; the functioning of the product in use; the re-use and/or recycling of the product; and appropriate and honestly informative advertising and sale – all to be within the bounds of sustainability.

It can be disheartening in the world of today, in which one gigantic crime after another is committed against the environment, to discuss sustainable design and manufacture. However many of the products shown here exemplify this, and later in Part I, Background Technology, are examples of sustainable energy sources such as hydroelectricity and wind power.

Learning about industrial design

One of the secrets of good design education is to ensure that all corners of the profession are experienced, and that the skills of the main trades of the built environment are introduced in a lively way to the student so that he or she really experiences the relationship and integration of the different design and manufacturing disciplines. Apart from ensuring a proper insight to the profession, it gives a good basis on which to find out where personal interests lie.

Anyone who has taught architecture and design knows that it is easier said than done. But, within the limited scope of this book, we have been able to place different disciplines and their skills side by side in an effort to encourage a broad integration of subjects and interests. One cannot learn and become qualified in all matters but one can create a picture for oneself of the fascinating range of materials, techniques and industries which we have at our disposal.

Common materials are so familiar to us that we think we know them – metal is cold to the touch and hard, wood is warmer and softer, and so on – and we tend to start from there, even in schools of architecture and design. But, of course, this is not sufficient. We need to know enough about materials, their properties, their origins and availability, to be able to take a critical attitude as to where and how we should use them in our design work. For example, all the materials in this book, with the exception of one (wood) are created under high temperatures. Two of these (metals and glass) are also worked hot, one of them (ceramics) is not. They consist of a mixture of materials; this mixture is often mentioned in the case of one of them – metal alloys. In the other two (ceramics and glass), this is seldom mentioned other than by a general name such as porcelain or crystal glass. Wood is the only one of the four materials which remains in its original form and composition and has nothing to do with heat when in its basic state; it is the only one which is organic and which we do not 'steal' from the earth – it starts as a plant and is replaceable. We can bend it with the help of heat but otherwise we work it entirely by cutting it in various ways. And so one could, and should, go on. But nobody ever says very much about all these absolutely basic facts and comparisons, so we generally do not develop the respect for materials which they deserve. Worst of all – but fortunately it is not so common as it has been – many students complete their training without ever having worked any materials with their own hands.

One of the greatest obstacles the student of architecture and design has to contend with is knowing how and why things are made in the way they are. Materials and their manufacturing techniques, and the way the components of a building or product are put together, are for the designer what colour, paint and brush are for the artist. The understanding of these complexities is one of the prerequisites for freedom to plan and design in a meaningful way. The words, drawings and photographs on these pages only get us some way towards an adequate understanding of design work. But the story they tell is the creative process in workshop, factory and drawing office.

We are here absorbed by the design-and-making synthesis but we should not forget that equally important dimension of architecture and design – the humanitarian. All the technical skill in the world does not make for good design if social engagement is lacking. Once we appreciate this professional duality, we then realise the actual infinity of design work. It is this double role, that the designer and architect have, that drives society.

What, in the context of this book, do we mean by 'industrial design', and with what criteria have we chosen artefacts to illustrate the subject? The simplest answer is that which has been designed for and made by industry, and the process of doing this. Nearly every item discussed is covered by this definition. The expression 'industrial design' does not mean a particular profession, style or way of doing things. Industrial design should not be something which is done simply to make things sell better – although this is how it is often regarded by industrialists, to the general detriment of the manufactured environment. Our criteria of choice have been both fairly liberal and discerning: objects which are formed, constructed and produced in a way to give people satisfaction in use, appearance and wear. Most of the items have an inevitable, intrinsic beauty. That is to say, it is an integral part of their form and function. Many of them have a certain 'musicality'. While working with each artefact they have given us some sort of lasting design input and we hope that they will do the same for the reader.

The educational value of the hand-produced, measured drawing is hard to exaggerate. When using this book, the drawing and photograph of each product should be studied together for complementary explanation. Most of the products are measured directly from the object to the paper and drawn in full size. This is the nearest the design student can get to 'making the object', without actually doing so. While measuring and drawing, which requires great accuracy and patience, one is allowing the object to explain the way it has been made, its exact sizes, shapes and materials. Where possible, especially for the smaller items of pottery and glass, drawings have been reproduced full-size so that measurements can be taken from them directly. But even where we have had to reduce the drawing because of the size of the object, the fact that the original drawing was full size makes for much greater accuracy. Where possible the drawing is reduced to an accepted scale; otherwise there is a graduated scale line. Here it should be mentioned, as a general point in design work, that where possible it is often an advantage to work in full size, rather than to scale, for the sake of accurate perception. All thicknesses, details, unevenness, asymmetry, signs of wear and use are faithfully recorded. As the reader can see from many of the drawings, this degree of accuracy tells us many significant, sometimes curious facts about the product and its manufacture. Any preconceived ideas about the perfection and accuracy of machine-made products are dispelled; asymmetry occurs frequently where symmetry was intended! In the same way that the product texts are not specifications, so the drawings are not working drawings. It is impossible to make either specifications or working drawings from finished products. The specification and the working drawing are instructions for manufacture and express the designer's intention; the measured drawing tells the actual truth about one particular finished product. There are exceptions; certain details – such as a dowelled joint between the rail and leg of a chair – cannot be recorded, simply because they are hidden. No dimensions are shown on the drawings as they would be if they were working drawings, intended for giving instructions. However, we have given the scale so that all dimensions can be scaled off, and usually one or more of the main dimensions are stated in the notes.

Apart from the reproduction of old etchings, the only other type of drawings are sketch measured drawings. These are much used in professional practice because they are quickly made, and are the method used if the object cannot be taken back to the drawing office. They are not nearly as accurate but can be dimensioned and sketched on for later use. Here we have redrawn them to scale.

A small number of drawings in the book are done by students as part of their studies at the School of Architecture in Aarhus. Measured drawings are discussed further in the introductions to Parts III and IV.

Our list of regrets

Many of the useful things of daily life are so ordinary to us that we take them for granted and do not notice their beauty and ingenuity. But often when we meet them in the context of a foreign country, as something new to us, we see their qualities. Obviously one cannot include everything one would like in a survey of this sort which tries to be both thorough and authentic. Difficult choices have had to be continually made between several equally qualified objects. Our list of regrets is long but the following are some of those we most regret:

To understand the meaning of the functional tradition in product design, there are few better places than the symphony orchestra and its instruments. Here is a concentration of superb designing and making by experience, tradition and skill in a multitude of materials, techniques and constructions designed to produce a great range of sound qualities. We are ashamed to say that we have no *musical instruments*.

Likewise the *camera*, which unlike the musical instrument, follows the development of industrialisation step-by-step, from being the very finest example of cabinet and instrument making to being the product of advanced precision manufacturing.

There is no hand-held *sports equipment* like the tennis racket or the cricket bat. These embody a fund of knowledge about strength, resilience, lightness and balance. The nearest we come to them is the wooden ski.

Folding rulers of box wood and brass we have had to leave out. They used to be the first items Danish students of architecture tried to buy when on study trips to England.

Ships and boats: the design of the traditional boat of the Faroe Islands is unique. They have the lines of the Viking ships, but are much smaller. Their form has evolved from the vital task they have to fulfil of speed, with safety, at sea. So they possess that special beauty resulting from the unity of function, form and construction. In the present collection, all the artefacts are land bound.

Like the boat, the *horse drawn coach* and the *farmer's cart* of the eighteenth and nineteenth centuries represent an important stage in industrial design which was based on a closely integrated co-operation between wood and iron, and so between coach builder, wheelwright and blacksmith.

Although it is very ancient, the *barrel* remains one of the key products of industrial design history. It exploits the strength and water-tightness of curved wooden staves when wet and constrained by iron hoops. This is achieved by a structural system in which the liquid content of the barrel is one of the component parts. It was too early to be included!

Nearer to our own time the *angle poise lamp*, which has made life and work possible in drawing office and factory for 50 years now, is not present, or the Swedish *Bahco shifting spanner*, or the American machine hand tool which – whether electric or petrol-driven – never seems to give up.

The well-known and much loved little *wooden chair*, J 39, which the Danish architect Børge Mogensen designed for FDB (the Danish Co-op) in 1947 and which has been in production ever since. This chair is friendly, light, strong, simple and timeless, and to be found in ancient churches, universities, and canteens alike.

The square *book case* with the oblong book compartments, which when one way up enables the storage of high books, and when the other way up, small books. The book case units measure only 76 x 76 cm and can be built up and arranged according to the book sizes in the owner's library. They were designed in 1928 by the Danish architect Mogens Koch, and have been beautifully made, ever since, by the Master Cabinetmaker Rudolph Rasmussen.

INTRODUCTION

Even something as increasingly treasured as the *wood and natural bristle brush* is absent. Likewise the English *garden tool* and the 'Wilkinson Sword' classics are regrettably absent. There is only one representative of the important design field of *street furniture*, street signing. Here we would also have liked to include the Paris *park chair* and the Jørgen Utzon *public telephone box*.

There is no representative of *ceramic building materials*. A chimney pot found in Sweden had to be omitted at a late stage.

No *optical glass* or spectacles, and at the opposite end of the glass industry, no cast pint and half pint *beer mugs*.

None of the good *household machines* from, for example, Miele, Nilfisk or Husqvarna. This is because we have had to make a policy of keeping clear of composite artefacts – except in Part I. In this part we would have liked, amongst many other things, the TE 20 *Ferguson tractor* of 1946, and wonders of finesse such as the *ball race* and *roller race*, to have been present.

The nature of materials

This is a huge subject which is discussed with the products that follow, but here are a few general points of interest.

The four materials, or material groups, represented here (together with textiles, papers, and plastics, which it is hoped will be the subject of a second volume) are the most used in manufacture. Only one of these, wood, is taken straight from nature. All the rest are the product of man's ingenuity and basically pre-industrial inventions, except one, plastics, which originated in the early 20th century. Natural materials such as bone, leather and stone are not included because, important though they are, their use in product manufacture is limited and wood is the only one of the natural group which has developed its own advanced form of mechanisation.

This choice of materials is in itself an industrial statement of significance: we deal nowadays almost solely with synthetic materials and it is difficult for us, in these climes, to imagine the time when materials such as bone, horn, skin, leather, cork, wood, clay, basket work, rattan, silk and wool – all hand worked – characterised the environment.

Every material has its own set of physical and manufacturing characteristics: degrees of expansion and contraction under various conditions of temperature and moisture, corrodibility, toughness, brittleness, structural make up, grained or homogeneous. These decide to a large extent the feel, strength and appearance of the product in which they are used. When several materials are involved in a product, these characteristics are of particular importance. Some materials, such as most metals, many types of yarn, wood and certain plastics, are ideally suited for working with each other and with other materials. Others, such as glass and earthenware are normally used alone, or as separate parts, and the design task is different. The simplicity of most of the products in this book is partly due to a choice of examples which demonstrate materials and their manufacturing techniques most clearly.

Each material category consists of a whole range of different types: types of metal, types of wood, types of ceramic materials and so on. This is one of the things which makes the study of materials so fascinating. By dividing metals into the two main groups, ferrous and non-ferrous, it helps to understand that every metal type in these two groups is an alloy (mixture) of several metals which have been composed through history; by experiment and experience as to which alloy is suited to (a) the manufacturing processes to which it is to be subjected and (b) the various functions and conditions of the finished product in use. The same in principle applies to all materials. Even wood varies from the softest to the hardest, from light to heavy, and from paleness to darkness according to the species of tree it comes from and the way in which it is cut. Many of these material variants are discussed in their respective chapters; likewise with forming techniques for the various materials and their surface finishes.

Materials also have general characteristics which apply, with variations, to most of them. With the exception of textiles and wood, a large proportion of them are castable, that is to say they can be formed in a mould when in a soft or liquid condition and become an exact replica of the interior of the mould, but in positive form. Casting and pressing techniques form the core of industrial processes and have been steadily refined for the various materials right through history. The latest material to arrive on the scene, plastics, has proved to be so eminently suited to casting that a very large proportion of the objects and component parts with which we are now surrounded are cast-made. The technique has been developed to such a pitch that with it we can control details such as surface texture and degree of shine, by the design and making of the moulds in which products are cast.

One of the curses of industrialisation is the imitation of natural materials with synthetic materials. The practice dates from the early 19th century and has accelerated ever since, with the plastics industry now taking pride of place and surrounding us with white lies. The plastics industry can deliver every imaginable material from 'leather' (sometimes even with stitching!) to 'stone', and 'wood' in all shades and grains. Functional standards have also been lowered by the indiscriminate use of plastic instead of wood for tool handles (wood is the best to hold in the hand), plastic instead of natural bristles for brushes and instead of wood or cloth for chair seats and backs. A closely related industrial disease has been the manufacture of items in inadequate, substitute materials because they are a little cheaper than if made in the correct materials: for years motorists had to endure car door handles, and other parts, made of chromium plated zinc alloy which broke easily and spot-corroded in the course of a few months. There are many examples of the ways misleading production and use of materials have impoverished the environment.

The nature and act of designing

It can be argued that society would have been better off if the expression 'designer' had never been invented – 'architect' covers the function much better. This idea is prevalent in non-English speaking countries partly because 'designer', though now adopted, is a foreign word, and also because it is such a general expression that most people do not understand it. If they think they do, then it has to do with fashion in clothes. And it is rather a lightweight term for a whole profession. Some of the best designers on the continent of Europe have been architects. That is to say, they have studied in the design departments of schools of architecture, where there should be a universal approach to design involving a planning, technical, and cultural depth.

INTRODUCTION

The work of both the architect and the industrial designer is placed in the very middle of the triangle of art, science and technology, and this is the main reason for both its fascination to those who practise it, and its complexity, and – as can be seen from the contents of this book – its enormous scope.

What is good design? We have already looked at the numerous types of designer and divided them into five main categories. From these descriptions we can see that good design consists of many things. At one end of the scale is the everyday utensil, simple or complex, which makes one think to oneself 'this is nice and easy to deal with, how good of the designer to think of that detail, or that function'. Much of good design has to do with being considerate, that is examining other examples in the field (including historical ones if they exist) and observing their good and bad qualities. By observation, experiment and consultation we find the answers to how do people really use a knife, or pour from a jug, or use a door handle, and so on. One can often tell by how a product has been designed – planned, shaped, articulated – whether the designer has had sleepless nights trying to resolve conflicting requirements and bring them to a finished solution. This is common to all design, micro and macro. At the opposite end of the design scale could be a product resulting from an engineering or scientific research programme which is an integrated part of the design project. For example the electric light bulb (pp 50-51) or the water turbine (pp 56-57), or the IC3 train (pp 70-71). This type of project is, like building design, normally the result of close co-operation between the members of a team of specialists in the field, technicians and designers.

David Pye, in his absorbing book *The Nature and Art of Workmanship* wrote:

> Our environment in its visible aspect owes far more to workmanship than we realise. There is in the man-made world a whole domain of quality which is not the result of design and owes little to the designer. On the contrary, indeed, the designer is deep in its debt, for every card in his hand was put there originally by the workman. No architect could specify ashlar (dressed stone – *ed.*) until a mason had perfected it and shown him that it could be done. Designers have only been able to exist by exploiting what workmen have evolved or invented.

This is a provocative statement. Can it be true? On the following pages we have searched for the answer.

PART I

BACKGROUND TECHNOLOGY AND INNOVATION

This book really consists of five 'volumes' which are meant to be used like a small library – not necessarily in sequence, but according to the subject, or subjects, in which the reader is currently interested. This is why the 'volumes' are called 'parts' rather than 'chapters'. Studying the orientation chart at the end of the book should clarify this structure. Here you will see all the pages at once – although simplified and in mini format. You will also see that all the parts of the book are closely related by being on a common chronological grid: they are each composed of double page spreads which correspond to the same decades (or, for the 18th century, not decades, but three periods).

Part I is the only section which does not deal with the products of a specific material. It is a service section to which one can return when one wants, for example, to place a product in its industrial and cultural context, or to see what else a designer or manufacturer has achieved, or if one simply wants to find information about a particular invention or event.

Each double page in Part I consists of a list of historical events of the period and a short, illustrated article about one of these events which is particularly relevant to the history of industrial design. These lists are in no way exhaustive. The entries are very brief, and are chosen for the picture they give to the designer of the way in which manufacture, energy forms, transport, communications, building and, that which is common to them all, *design,* have developed. They are based on findings we have made during our product research, and items selected from the lists of others. Here we are specially indebted to the work of Lewis Mumford, and T.K. Derry and T.I. Williams.

These events are not only some of the facts of industrial development: they also tell us something about the stages of industrial creation. This often follows a course which starts with the very first idea to achieve, or improve something. Some years later a re-evaluation and re-design is introduced, and the process can be repeated several times by different people at different periods with improvements on each occasion. So, when using the events lists, it helps to understand the character of invention: the gradual, historical maturing from idea to solution. Total invention seldom happens suddenly. The typewriter is a typical example of this maturing process. Under the continual influence of both changing human needs and changing technical possibilities its basic design has been under change and development from 1714 to the computer solution we know today.

BACKGROUND TECHNOLOGY AND INNOVATION

Starting at the beginning of the 18th century we are coming into the history of technology at a comparatively late stage. Mechanisation had just started to become evident and with it the beginnings of a shift from a predominately wood technology to one of metals. But most of the spadework – the basics – had already been achieved. The clock, the lathe, the loom and the printing press were established. The advanced engineering of watermill and windmill had long adapted them to produce a wide range of essentials.

In order to give structure and form to the complex and devious route of technology as shown on the pages of Part I, we would like to mention here some of its higher peaks, and sharper bends. There have been remarkable constellations of industrial craftsmen, as well as particular individuals, who have changed manufacturing practice. Accomplishments and events where technology and design have come very close together, often unnoticed, but with superb results.

The craftsman-engineers. The lock, the pulley block machines, the steam-hammer and the standardisation of the screw are, in a sense, all linked. Between 1790 and 1840 their inventors, who were some of the first highly skilled and dedicated industrial craftsmen, changed the face of mechanical engineering from being relatively crude to being something in which extreme precision and accuracy mattered. Joseph Bramah, the inventor of the hydraulic press and the Bramah lock, led the way with what has been described as the first proper machine shop. Henry Maudslay was employed in Bramah's workshop before he started his own. It was he who developed and made the Portsmouth blockmaking machinery designed by Marc I. Brunel. It was Maudslay's insistence on previously unknown precision in machine tool making which inspired his employees and put machine manufacture on a new qualitative level. Amongst his employees were James Nasmyth, inventor and maker of the steam hammer, and Joseph Whitworth, machine tool designer and manufacturer, who initiated the standard screw thread which carries his name.

The American system. At the beginnings of industrialisation, there were a number of similar constellations of brilliant craftsman-engineers, not least in USA. Here the cause and effect of mechanisation was very different from that of Europe. It started at the very beginning of the 19th century, with some urgency, as a substitute for, and in the abscence of, skilled labour. The machine did not cause the same disruption and misery to start with as it did in Britain, and was soon accepted and became the norm. Here there was a quicker and more vigorous start to mechanised production with necessities like tools, clocks, and sewing machines acting as 'locomotives', and later on harvesters, typewriters and many other machines. Mechanisation *had* to work and it had to produce the goods in very large numbers. The key to all this was the adoption of the practice of interchangeability of the component parts of products. At the time this was known, rather quaintly, as 'The American System', but ever since has been seen as nothing less than a fundamental principle of mass production - so much so that nobody talks about it any more. In short it means that during assembly any part (for example a cogwheel) must fit together and work with any other related part (its axle) after leaving the machine tool, without prior hand fitting.

Standardisation. This brings us to the whole question of standardisation. In many respects standardisation is the natural extention of interchangeability, but it is concerned with many aspects of a product. It is often a form of guarantee specifying quality of materials, dimensions, design, etc. of components and the

26

finished article, for the benefit and control of the producer, and for the protection of the consumer. Standardisation brings a minimum of order into an uneven production and marketing system. It is made necessary by, and is fundamental to, a civilised industrial society. One of the anomalies of industrial history is that standards legislation was established as late as the early twentieth century. One can say that the necessity to standardise reflects both the best (the assertion of quality) and the worst (unethical industrial practices) in the industrial society.

The cock-crow of triumph. Standing right in the middle of our industrial story is the Crystal Palace. Lewis Mumford wrote:

> The phase one here defines as paleotechnic reached its highest point, in terms of its own concepts and ends, in England in the middle of the nineteenth century: Its cock-crow of triumph was the great industrial exhibition in the new Crystal Palace at Hyde Park in 1851: The first World Exposition, an apparent victory for free trade, free enterprise, free invention, and free access to all the world's markets by the country that boasted already that it was the workshop of the world.

The numerous international trade exhibitions, held in various capitals of the world, which followed, are recorded in the events lists under Architecture and Design.

A designer-manufacturer. Most of the giants of early industrialisation have been men of metal. It was the mastering of metals which enabled mechanisation. But equally important to the study of industrial design are all the other materials. Until the industrial revolution wood had been the central material, and it is natural – and telling – that our first article in Part I is concerned with engineering in wood. But wood crops up again much later with the extraordinary achievements of Michael Thonet and his design and manufacture of chairs in steam bent beech wood. Thonet's chairs are still on the market – an honour he shares with Joseph Bramah and his locks.

Most of the subjects touched on in this introduction are discussed in greater detail and illustrated in Part I. A few appear in the events lists only but are taken up in the other Parts.

PART I. BACKGROUND TECHNOLOGY AND INNOVATION 1700-1750

Entries marked with an asterisk are discussed in greater detail

Manufacturing
Sheet glass by pouring and rolling, de Nehou 1700
Wet sand iron casting, Darby I 1708
Clockmaker's precision lathe of metal *c.* 1705
Salt glaze for ceramics spreads from Delft 1710s
Coke used in blast furnace instead of charcoal, Darby I 1709
Hard paste porcelain starts in Europe, Böttger 1710
Sewing machine, De Camus 1711
Mercury thermometer, Fahrenheit 1714
White salt glaze pottery, Astbury, *c.*1720
Underwater flint grinding for china 1726
Plant fibres begin to replace rags for papermaking, Réamur 1732
Flying shuttle, Kay 1733
Roller spinning, Wyatt & Paul 1733
Power plate rolling machine, Polhem 1734*
Gear cutting machine, Hindley, *c.*1740
Casting pots in moulds begins to replace hand turning
Transfer printing of pots begins to replace hand decoration 1740s
Mechanical loom for figured silks, Vaucanson *c.*1740
Crucible steel, Huntsman 1745

Energy
Water-power for mass production, Polhem 1700*
Steam pump, 'the miners friend', Savery 1702
Atmospheric steam engine, Newcomen 1703

Transport
World navigation chart, Halley 1702
Wooden railways covered with iron 1716
Chronometer no.1, Harrison 1736
Cast iron rail tramway, Whitehaven (England) 1738
Theoretical structure studies on hull strains at sea, Bouguer 1746
Scientific calculation of water resistance to hulls, Euler 1749

Communications
Stereotype, Van der Mey & Müller 1710
Typewriter, Henry Mill 1714
Three-colour printing from copperplate, Le Blond 1719
Stereotyping process, Goldsmith 1730

Architecture and Design
St. Petersburg founded. Leading architect Tressini, 1703
The mechanical laboratory and the Mechanical Alphabet, Polhem*
St. Paul's Cathedral, London, Wren 1711
Nyboder seamen's housing, Copenhagen 1648-1795

Society
Exact measurement of blood pressure, Hales 1727
Handel's Messiah first performed, Dublin 1742
First technical school divided from military engineering, at Braunschweig 1745

Names
Christopher Wren 1632-1723
Thomas Savery 1650(?)-1715
Christopher Polhem 1661-1751*
Thomas Newcomen 1663-1729
Abraham Darby I 1678-1729
Gabriel Daniel Fahrenheit 1686-1736
Benjamin Huntsman 1704-1776

Christopher Polhem
The work of Christopher Polhem was concerned with industry and with mechanical design and, therefore, is eminently suited for starting this study of the historical development of industrial design. It can, of course, be argued that the development of the steam engine was by far the most important single activity throughout this period, and so it turned out to be. But it is of interest to consider that important preparations for industrialisation were also taking place outside Britain. Polhem was a Swedish industrialist, inventor and teacher of mechanics. His accomplishments spanned from the design of machine tools to the water-power plants which drove them. The machine, 'A', made brass cogwheels for clocks and watches in several sizes simultaneously and could be operated by one man. It is believed to be one of the world's first proper machine tools. Of equal interest to Polhem's production machines are his models of basic mechanical movements, which he made and used for the teaching of mechanical design between 1696 and 1735 in what came to be known as the Mechanical Laboratory. This systematised collection of machine movements he described as the 'Mechanical Alphabet' and ultimately, with the help of students, the alphabet reached 103 letters', or models, three of which are shown here.

A. *Machine for making clock parts 1729*

Three 'letters' from the Mechanical Alphabet:

B. *With a back and forth movement of the arm the wheel is given continuous rotation, with the help of two spring-loaded racks which engage alternately in the hub wheel.*

C. *Various cam mechanisms.*

D. *The top left mechanism shows how a pendulum movement of the long arm is converted to a rotary movement of the ratchet wheel. Top right shows how an escapement actuated by a pendulum, at the rear of the panel, gives a rotary motion in the ratchet wheel. Bottom left shows a gear ratio. Bottom right a ratchet wheel with spring loaded pawl.*

Museum of Technology, Stockholm
Photos: Rigmor Söderberg

A

B

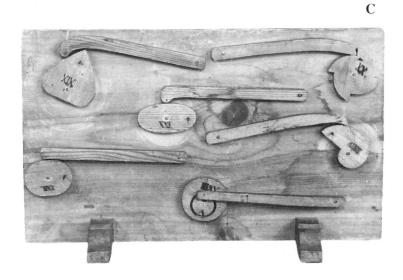

C

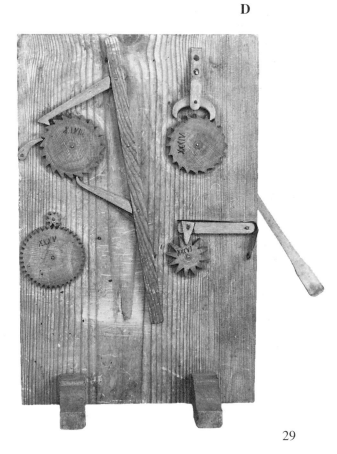

D

Entries marked with an asterisk are discussed in greater detail

Manufacturing
Bi-metallic curb in Harrison's chronometer 1750s
Bone china 1750
Double firing of ceramic products, Booth 1750
The influence of carbon content on the hardness of steel, Bergmann 1750
Isolation of nickel, A.F. Cronstedt 1751
Advanced silk factory near Lyons, Vaucanson 1756
Wedgwood's pottery firm founded 1759
Prismatic lathe-bench, replacement of wood by metal for lathe construction 1760s Small drill, Jacque de Vaucanson 1760s
Piston and cylinder air pump (water wheel driven) tripled blast-furnace production, Smeaton 1761
Metal lathe slide-rest 1763*
Soho Foundry completed, Boulton,1766
Spinning Jenny, Hargreaves 1767
Small screw-cutting lathe, Ramsden 1770
Heavy engineering lathe, Vaucanson 1770s
Process for manufacturing smaller articles in papier mâché, Clay 1772
Boring mill for cannon (leading to cylinders for steam engines), Wilkinson 1775
The Royal Danish Porcelain Factory 1779

Energy
First two James Watt steam engines installed 1774

Transport
Iron wheels for coal cars 1755
Chronometer No 4, Harrison 1759
Canal building starts in England, Brindley 1760s
Captain Cook's ship *Endeavour* to South Pacific 1760s
Improved stone road construction,Trésaquets 1764
Cast iron train rails, Coalbrookdale 1767
Steam carriage, Cuquot 1769
Caterpillar tread, Edgeworth 1770
Copper sheathing for ships bottoms 1770s
Iron ribs and reinforcing in wooden ships 1770s

Communications
Baskerville type, Baskerville 1757

Architecture and Design
The Academy of Fine Arts (architecture, sculpture, painting) founded in Copenhagen 1754
Amalienborg Palace, Copenhagen, Eigtved 1754
The Gentleman & Cabinetmaker's Director, Chippendale 1754
Cement manufacture, Smeaton 1756
Stone lighthouse at Eddystone, Smeaton 1760
Arkwright's first textile mill at Cromford 1771
Bedford Square, London, Thomas Leverton *c*.1775
Piece Hall Cloth Market at Halifax, England 1775
First 'modern' type water closet, Bramah 1778

Society
Ludvig Holberg dies (1684-1754)
Encyclopédie des Sciences, des Arts et des Métiers, Diderot 1751-1777*
First exhibition of industrial arts in Paris 1763
Encyclopedia Britannica, England 1768
Royal Agricultural Society of Denmark 1769
American Declaration of Independence 1776
The Wealth of Nations by Adam Smith 1776
Christiansfeld, Denmark, founded by the United Brethren (Herrnhuter) 1772
W.A.Mozart (1756-1791) at his prime

Names
John Baskerville 1707-75
Jacques de Vaucanson 1709-1782
Denis Diderot 1713-1784*
James Brindley 1716-1772
Thomas Chippendale 1718-1779
Adam Smith 1723-1790
John Smeaton 1724-1792
John Wilkinson 1728-1808
Mathew Boulton 1728-1809
Josiah Wedgwood 1730-1795
James Watt 1736-1819

The slide rest
The slide rest is part of the lathe. It is a robust holder for the turning tool which can be slowly and accurately moved both towards and away from the work as well as parallel and obliquely to it. Its development accompanied that of the lathe when this ceased to be a wooden machine and became one of metal, first for clock making, later for the turning of heavier machine components and screw threading of metals. The three oldest machine tools, the potter's wheel, the drill and the lathe, are those which came long before industrialisation and yet were, and always have been, fundamental to it. The lathe and the drill enabled another basic requirement, the machine screw. These inventions gave man the ability and understanding to visualise mechanisation on a new level. They were necessary for the design and making of the steam engine and were later to be driven by it. The potential power of steam could only be realised by precision machining. Bearings, cylinders and pistons had to fit exactly and be perfectly round and the achievement of these rudimentary requirements of the efficient machine had always been evasive. The slide rest was the solution to this precision. In the way it held and allowed the guiding of the cutting tool, it was the secret to the vibration-free working of metal.

Henry Maudslay is often credited with the invention of the slide rest, but as with so many inventions of this productive period, others were working in the same direction. This illustration from the great French encyclopaedia by Diderot shows one of the very first slide rests. The similarities with the modern one shown below are astounding.

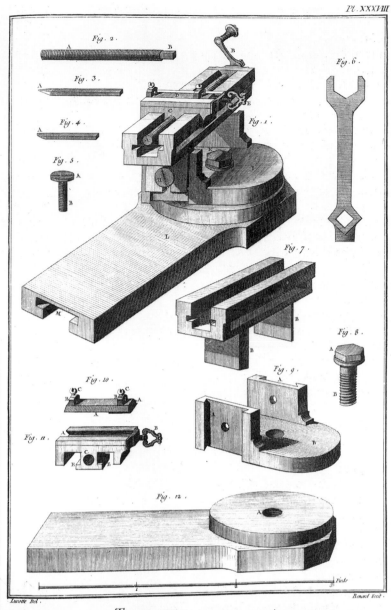

Tourneur, Suports Composés.

Small general-purpose lathe with slide rest mounted, Coronet 1953.

Entries marked with an asterisk are discussed in greater detail

Manufacturing

Machine tools for making lock parts, Bramah 1780
Metal sawing machine, Bramah 1780s
Rollers for manufacturing iron bars, Cort 1783
Puddling process or reverberatory furnace, Cort 1784
Spinning Mule, Crompton 1784
Interchangeable parts for muskets, Le Blanc 1785
Power loom, Cartwright 1785
First steam spinning mill, at Papplewick, 1785
First steam flour mill, Albion Mill, London 1786
Bleaching and dyeing of textiles, Bertollet 1780s
Dividing engine, Henri Gambey 1790s
100% metal machines, Maudslay 1790s
Bench micrometer 1790s
Heavy vertical boring machine, Boulton & Watt 1790s
Wood planing machine, Bramah 1790s
Feldspar added to bone ash producing the easily worked English porclain of today, Spode 1790
Lead pipe drawing machine, Wilkinson 1790
Sewing machine first patented, Saint 1790
Cotton gin, Whitney 1793
Rope making machine patented, Huddart 1793
Food preservation by bottling, Appert 1795
Hydraulic press, Bramah 1796
Large screw cutting lathe, Maudslay 1797
Enamelling of iron hollow ware patented 1799

Energy

Steam engine established as prime mover 1780s
First steam engine installed at Wedgwood's factory, Boulton & Watt 1782
Coal gas for lighting, Murdock 1792
Hydraulic ram for pumping water, Montgolfier 1797
Continual electrical current, Volta 1799

Transport

Introduction of iron chain cables and rigging for ships 1780s
Hot air balloon, J.& E. Montgolfier 1783
Hydrogen balloon, Charles & Robert 1783
Steam carriage, Murdock 1783
Steam boat on Saône, Joufroy 1783
Barge of bolted cast iron plates, Wilkinson 1787
Screw propeller steam boat, Fitch 1787
Road construction, McAdam *c*.1795
Ball bearing for vehicle axle, Vaughan 1794
Canal building at its height in England 1790s

Communications

Oil lamp with cylindrical burner and glass chimney, Argand 1784-5
Semaphore, Chappe 1793
Lithography, Senefelder 1796
Optical glass, Guinard 1798
Paper-making machine, N.L. Robert 1799

Architecture and Design

W.C. with water seal patented, Gaillait 1782
Masson Mill (cotton spinning), Arkwright 1784
Lock mechanism patented, Bramah 1784
New Lanark industrial village, first stage, Dale & Arkwright c.1785
Soho Foundry, Boulton, Watt & Murdock re-built and extended 1796
First iron framed factory (cast iron), C. Bage 1796
Iron Bridge, Darby III and T.F. Pritchard 1779 *

Society

First recorded kitchen cooking range, Robinson 1780
The Times, London, founded 1785
Shaker community founded at New Lebanon (later Mount Lebanon) USA 1788
The French Revolution 1789
The metre defined by the French Academy of Sciences 1789

Names

Richard Arkwright 1732-1792
Henry Cort 1740-1800
A.G.A.Volta 1745-1827
Joseph Bramah 1748-1814
Abraham Darby III 1750-1789 *
Samuel Crompton 1753-1827
Aimé Argand 1755-1803
J.L. McAdam 1756-1836

The Iron Bridge

The Iron Bridge was designed by the architect Thomas Pritchard and the ironmaster Abraham Darby III. It is the first bridge to be built entirely of iron and the only one of its size from this period which is still standing. It is situated at Ironbridge, the small town which takes its name from the bridge, in a gorge of the river Severn in Shropshire, England. The Ironbridge Gorge, which is now an industrial museum region, was one of the very first centres of industrialisation. The bridge was erected between November 1777 and January 1781. It has a span of about 30.7 m and a height of about 12.7 m, both measurements taken at towpath level. Its width at road level is about 7.5 m. It is constructed of cast iron parts made at Darby's ironworks and then transported, erected and assembled on site. With this new structural technique, the bridge signalled the first beginnings of the prefabrication of large structures, here necessitated by the need for a large clear span to allow passage of heavy river traffic on a site with no room for gradually rising approaches.

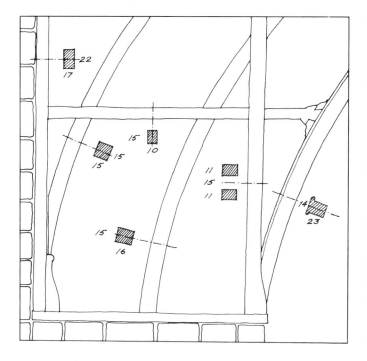

A measured sketch and a photograph both taken from the towpath showing the large sections and carpenter-like jointing with wedges, dowels and dovetails. Dimensions given in cm.

Entries marked with an asterisk are discussed in greater detail

Manufacturing

Du Pont de Nemours & Co. 1802
Woodworking machinery for making pulley blocks, M.I. Brunel & Maudslay 1803 *
Interchangeable parts for muskets, Whitney
Machine dresser for cotton warps (necessary for power weaving)
Loom for figured fabrics, Jacquard 1804
Knitting machine, Heathcoat 1809
Canned food, Durand 1810

Energy

Galvanic cell, Volta 1800
High pressure steam engine, Trevithick 1803
Soho Foundry illuminated by gas, Murdock 1804

Transport

Steam road carriage, Trevithick 1801
Horse drawn public railway, South London 1801
Steamboat *Charlotte Dundas*, Symington 1801
Railway locomotive, Trevithick 1804
Twin screw propeller, Stevens 1805
Gas driven motor car, Isaac de Rivaz 1807
Side-paddle steamboat *Clermont* Fulton 1807
Göta Canal started, Telford 1809 (–1833)

Communications

Printing press of iron, Stanhope & Walker 1804

Architecture and design

Fire proof factory, Strutt 1804 *
Pontcysyllte canal aqueduct, Wales, Telford 1805
Suhr's & Ludvigsen's warehouse, Copenhagen 1805

Society

Metric system established, France 1801

Names

Alessandro Volta 1745-1827
William Strutt 1756-1830 *
Thomas Telford 1757-1834
Eli Whitney 1765-1825
Marc Isambard Brunel 1769-1849 *
Richard Trevithick 1771-1833
Henry Maudslay 1771-1831 *

The Portsmouth blockmaking machinery

The machines were designed by the engineer Marc Isambard Brunel in co-operation with the maker, Henry Maudslay, to speed up the production of ships' pulley blocks for the British Navy. Production by hand had become a serious bottleneck in the dockyards. The demand for blocks had become enormous and by 1808 the new machines had an annual output of 130,000 blocks. The plant comprised 22 different machines. Some of the machines types were repeated but in the different dimensions necessary to make a wide range of different block sizes – from 4 in. to 18 in. long – each with from 1 – 4 sheaves, so that in all some 45 machines were involved. Historically these machines are remarkable. Nothing like them had been made before. They were for the first time all metal: cast iron, wrought iron, bronze and steel. Not even the supporting frames were made of wood. They were the first machine tools for quantity production on such a scale. Historically they can be seen as a technological catalyst: they inspired and influenced subsequent engineering practice and set new standards. The totally new manufacturing method left its previously handmade product basically unchanged, but the machines themselves are a classic example of brilliant conception, design, product development, manufacture and finish – qualities which meant that many of the machines were in regular use 150 years later. They are of particular interest to the industrial designer because they are the original 'prototypes' of today's woodworking machines incorporating most of their machining principles. See Pulley Block, pp 130-131

Illustration from 'The Cyclopaedia of Arts, Sciences and Literature' by A. Rees, 1819.

4 different saws for converting timber to regular blocks	boring machines for making axle holes and start holes for mortiser	mortising machines for making sheave holes	saws for removing corners from blocks	shaping machine for rounding the outer form	scoring machine to make the strop grooves

The shell   *illustrated*   *illustrated*

The principal machines and processes in the production lines   **assembly**

The sheave

cross-cut swing saws for slicing tree trunk into disks	cylindrical saws for rounding the discs and boring holes for bearings	routers for recessing sheave faces to take bearing lugs	drilling and riveting machines for riveting bearing lugs	broaching machine for centring and smoothing the bearings	lathe and polishing machines for finishing the forged pins

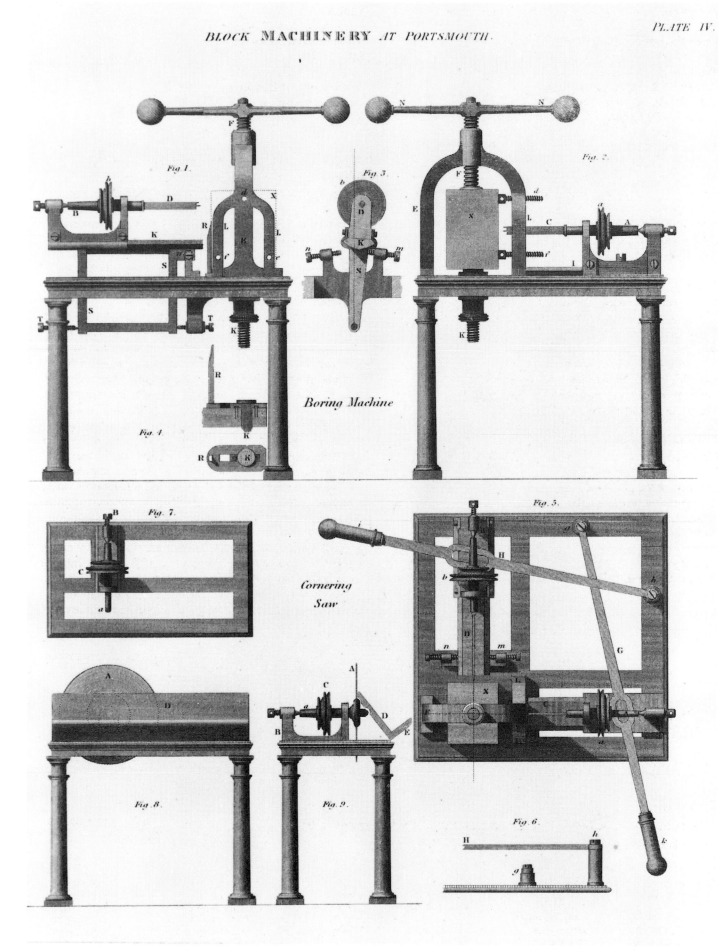

BLOCK MACHINERY AT PORTSMOUTH.

PLATE IV.

Fig. 1.

Fig. 3.

Fig. 2.

Boring Machine

Fig. 4.

Fig. 7.

Fig. 5.

Cornering
Saw

Fig. 8.

Fig. 9.

Fig. 6.

Manufacturing

Interchangeability (or the American System) starts in this decade based on the work of: 1) Le Blanc (Fr.) 1785, 2) Whitney (USA) 1798, 3) Bodmer (Germ.) 1806, 4) John Hall (USA) 1811

Power loom, Horrocks, 1813

Spinning of metal to form hollow ware etc., introduced in Paris 1816

Planing machine for metal, Roberts 1817

Cadmium discovered 1817

Milling machine, Whitney 1818

Energy

Westminster streets lit by gas 1814

Transport

Steamship *Comet* 1812

Locomotive *Puffing Billy*, Hedley 1813

Push bicycle, Drais 1818

Communication

Steam-powered cylinder printing press, Koenig 1814

Incandescent lamp, De la Rue 1820

Architecture and Design

Waterloo Bridge over Thames, Rennie 1817

Wrought iron glazing bar enabling large glass structures, Loudon 1817

New Lanark industrial community, Owen 1818

Society

The Napoleonic Wars 1805-1815

Names

John Rennie 1761-1821

Robert Owen 1771-1858

John Claudius Loudon 1783-1843

The cotton mill

This was the environment of the early adventures of factory mechanisation – the repetitive machine-manufacture of the repetitive product, cloth. It all looks very serene and ordered on the drawing but in fact was dominated by the deafening racket of batting, carding and spinning machines. Coal, iron and steampower are usually regarded as the catalysts of the Industrial Revolution, but it was waterpower which showed the way to the factory system, and never more impressively than as prime mover for the numerous machines invented for the preparation and spinning of cotton fibres. It was this new imported material which because of its strength, elasticity and cheapness stimulated the mechanisation of the textile industry in mills like the one shown here. The mechanisation of weaving had to wait for the consistent strength and quality of machine-made yarns and, to start with, this mill was still selling its yarn production to hand weavers in their cottages. At the weaver, the yarns were united with wool or linen and the consistent quality of mechanised spinning also made possible the manufacture of all-cotton cloths for the first time.

This spinning mill was typical of many though, as we shall see, the construction of the building was epoch-making. About 14 different manufacturing processes were carried out, most of which were done by closely attended machines. First came the unpacking and inspection of the bales of cotton, then several processes for cleaning, blending and classifying. This was followed by straightening and separating the fibres by carding, then finish carding, followed by drawing and roving to a loose, slightly twisted rope. Finally came spinning to produce yarn of the required quality, strength and consistency according to final use. The section fig.1 shows drawing machines on the 4th floor, carding machines on the 3rd and 2nd floors and spinning machines on the 1st and ground floors. On the 5th floor is the school for the child workers. All these – some 260 machines – were driven by a system of shafts and gears led up from the great water wheel which was 7 m in length and 5.5 m in diameter: the whole building was by way of being a giant machine. Also operated by the water-driven shafting was a lift serving the five factory floors and situated next to the staircase at P, fig 1. They called it the 'crane' and it was probably the first of its type.

This was the first mill to be built entirely of non-combustible materials. The owner, William Strutt, had suffered fires in other mills and became a pioneer in fireproof building construction. The columns (at 3.93 x 2.74 m centres) and cross beams were of cast iron, with shallow barrel vaulting of hollow, weight-saving bricks between. Wrought iron tie rods between the column tops resisted outward thrust. The floors were of brick. The roof structure was cast iron. Apart from being relatively fireproof this was a very solid construction which was suited to the motion and weight of the heavy, primarily iron machinery.

Illustration from 'The Cyclopaedia of Arts, Sciences and Literature' by A. Rees, 1819.

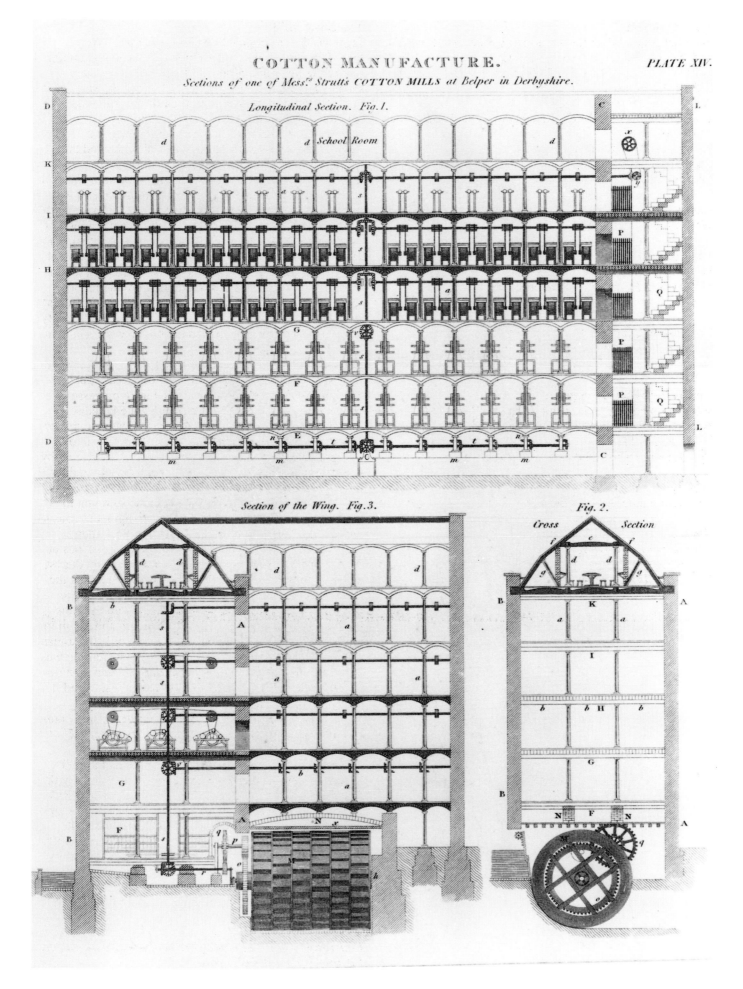

COTTON MANUFACTURE.

PLATE XIV.

Sections of one of Mess.rs Strutt's COTTON MILLS at Belper in Derbyshire.

Longitudinal Section. Fig.1.

d School Room

Section of the Wing. Fig.3.

Fig. 2.

Cross Section

Entries marked with an asterisk are discussed in greater detail

Manufacturing

Copying lathe to make gunstocks, axe handles, clog soles etc., Blanchard *

Steel alloys at laboratory stage, Faraday 1822

First factory for producing rubberised raincoats, Macintosh 1824

Thin sheet metal by rolling, *c*.1825

Tinplate, *c*.1825

Machine for making the lapped seam in products of tin-plate

The isolation of aluminium, Ørsted 1825

Pressed glass process, Bakewell 1825

Tongue and groove machine 1827

Reaping machine, Bell 1826

Scientific approach to strength of materials

Energy

Electromagnetism, Ørsted 1820

Principle of motor, Faraday 1823

52 English towns lit by gas by 1823

Electro-magnet, Sturgeon 1825

High pressure steam boiler (1400 lb) Perkins 1827

Transport

Iron steamship, Manby 1822

Stockton–Darlington railway opened 1825

Wire suspension road bridge, Sequin 1825

The steam locomotive *Rocket*, Stephenson 1829

Cammell Laird (Birkenhead) begin iron ship building 1829

Communications

Chromo-lithography, Zahn 1827

Blind print, Braille 1829

Paper matrix stereotype, Genoux 1829

Architecture and Design

Portland cement, Aspdin 1824

Thames tunnel, M.I. Brunel 1825-1843

Menai suspension bridge, Telford 1826

University of Virginia, Jefferson 1826

Technical Schools of Design & Technology start in Prussia 1827

St. Katharine's Dock, London, Telford 1828

Scientific approach to strength of materials

Society

First scientific congress, Leipzig 1822

Royal Life Boat Institution founded, UK 1824

Names

Thomas Jefferson 1743-1826

Hans Christian Ørsted 1777-1851

George Stephenson 1781-1848

Thomas Blanchard 1788-1864 *

Michael Faraday 1791-1867

The copying lathe

The copying lathe is a machine for turning wood to oval or eccentric shapes which it copies directly from a pattern. As with the Portsmouth blockmaking machinery previously discussed, the invention of the copying lathe was the solution to a problem posed by the armaments industry: in this case how to increase and cheapen the production of gun stocks. Again, as with the blockmaking machinery, the Blanchard machine was one of a series, each of which performed the different operations necessary to produce nearly finished gun stocks from the sawn wood. This earliest copying lathe introduced wood machining principles on which such lathes are still based: the fast rotating cutter (which does the shaping) which is linked and directed by a cam mechanism to a slowly turning pattern (which is being copied) and to the blank which is turning in unison with the pattern. The fact that the blank is being axially turned puts one in mind of the lathe but, unlike the lathe, the cutter tool is rotating. This is why the machine is also known as a copy moulder, or mill, in some countries.

The great design-historical significance of the copying lathe concerns its uses in the hand tool, chair-making and shoe industries. It did not only enable the quantity production of identical complicated forms in wood. R.A. Salaman, the tool historian, disputes the assumption that the well-formed axe handle as we know it from the middle of the 19th century is derived from its handmade predecessor. These were generally straight, or nearly so. It would therefore appear that Blanchard's invention itself led to the ergonomically sound handle form which is comfortable to hold and enables improved power and control of the axe. With this one exception the mechanisation of later years has led to a general impoverishment of the hand tool. The copying lathe made possible an improvement in handle design and has thus been one of the hand craftsman's few industrial allies.

Production stages of the clog sole: bottom, straight from the copying lathe – note the cutter marks. Middle, cutter-marks removed on sole edge. Top, all surfaces sanded smooth and rebated to take leather.

Oval sectioned handles of hickory for small axe, or hatchet, and for a carpenter's mallet.

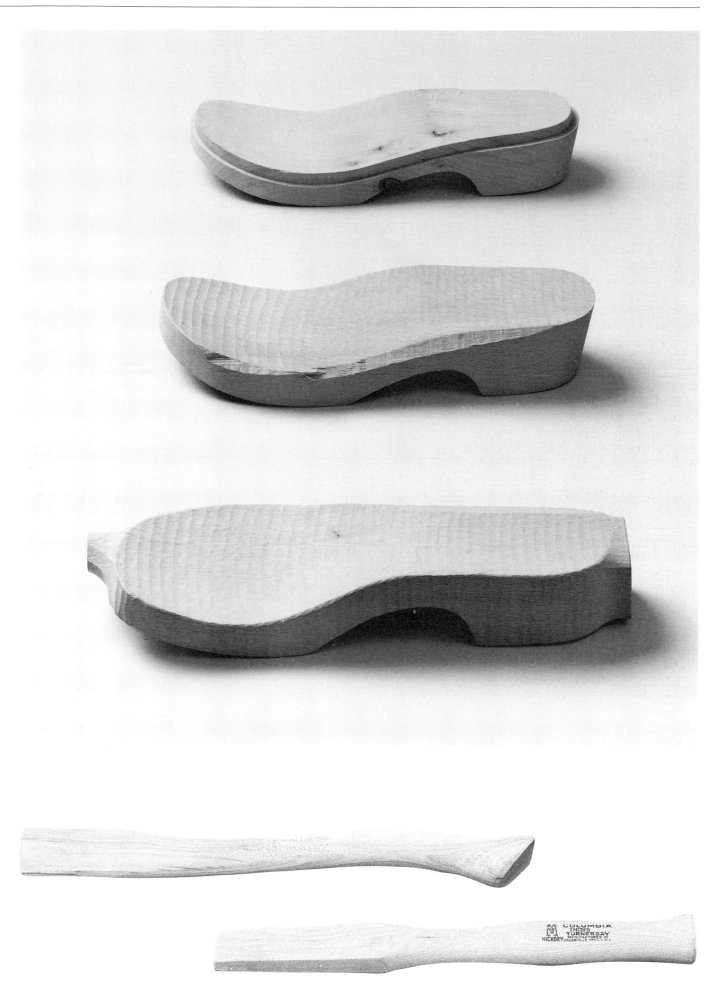

Entries marked with an asterisk are discussed in greater detail

Manufacturing
Self-acting spinning mule perfected, Roberts 1830
Reaping machine, McCormick 1831
Revolver pistol patented, Colt 1835
Galvanised iron, Sorel 1836
Patent Office established, USA 1836
Rubberised cloth, Hancock and Macintosh 1837
Steam hammer, Nasmyth 1839 *
Manganese steel, Heath 1839
Existence of vanadium, Sefström
Tunnel kiln for firing ceramics, Denmark 1839

Energy
Electromagnetic induction dynamo, Faraday 1831
Water turbine, Fourneyron 1832
Laws of electrolysis, Faraday 1833
Liquid refrigerating machine, Perkins 1834
Electric motor, Davenport 1835
Refrigeration by compressed air, Gerschel 1835

Transport
Liverpool–Manchester railway 1830. The railway era
    begins
Elevators begin to be used in factories
Electric battery in power boat, Jacobi 1834
Electric motor car, Davenport 1835
First use of electric telegraph on railways, Robert
    Stephenson 1836
Propeller steamship *Archimedes*, J. Ericsson 1838
First trans-Atlantic crossing under steam, *Sirius* and
    Great Western 1838
Clydeside build more iron ships than wooden 1839
The tea clippers sail Shanghai to London in 90 days
Göta Canal across Sweden completed

Communications
Magnetic telegraph, Gauss & Weber 1833
Electric telegraph 1835
Postage stamp, Rowland Hill 1837
Electro-magnetic telegraph, Morse 1838
Electrotype, Jacobi 1839
Daguerrotype photography, Daguerre 1839
Stearin becomes commercially available
Difference engine (seed of the computer), Babbage.

Architecture and Design
Regents Park development, London, Nash & Burton
Covent Garden market building, Fowler 1833
Industrial complex at Brede, Denmark 1832-1867
School of Design (later Royal College of Art),
    London 1837

Society
Chloroform, 1831
The word 'socialism' first used, France 1832
Application of statistical method to social phenomena,
    Quetelet 1835
Queen Victoria 1837
Peoples Charter (Chartism), England 1838

Names
John Nash 1752-1835
Thomas Hancock 1786-1865
Richard Roberts 1789-1864
L.J.M. Daguerre 1789-1851
Charles Babbage 1791-1871
Rowland Hill 1795-1879
Decimus Burton 1800-1881
Robert Stephenson 1803-1859
John Ericsson 1803-1889
James Nasmyth 1808-1890 *

The steam hammer designed by James Nasmyth in 1839 was a key invention without which heavy industry would have been unable to advance. It was designed to solve a particular problem. The steamship *Great Britain*, designed by I.K. Brunel, was initially to have had paddle wheels and Nasmyth was asked if he could forge the 76 cm diameter axles, existing tilt hammers being unable to take work of this size. Nasmyth's invention was the cue for the introduction of steam drop hammers for all sorts of uses where great impact force was needed. But die forging remains one of the most important functions of the drop hammer. Machine parts, tools etc., which have to endure stresses, can in this way be both formed and strengthened in the same operation. The steam hammer consisted of a great weight – the hammer head – connected to a piston which was raised by steam to various heights according to the force of blow required and then dropped onto the blank. One of the refinements of the invention was that the hammer could be given additional force by admitting steam above the piston to accelerate the downward stroke.

The photograph shows a working model.
The painting is by James Nasmyth himself.
British Crown copyright Science Museum London.

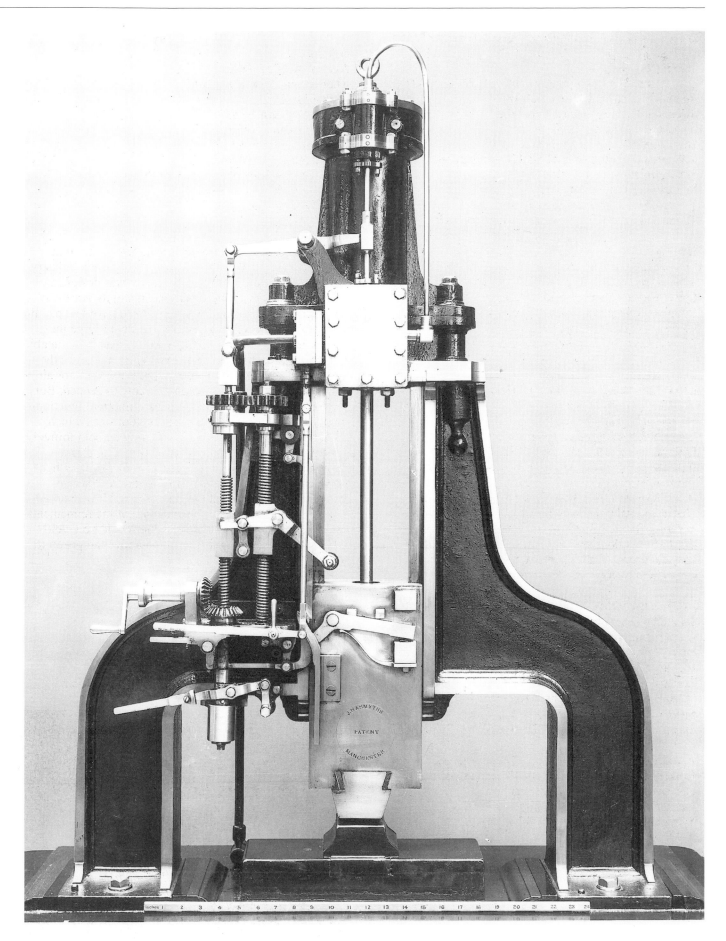

Entries marked with an asterisk are discussed in greater detail

Manufacturing
Electroplating of copper, brass with silver or gold, G.R.& H. Elkington 1840
Automatic spinning mule, Roberts 1840
Drilling machine, Nasmyth *c*.1840
Clay dust pressing for tile manufacture, Prosser & Minton 1840
Standard screw sizes and threads, Whitworth 1841*
Hot vulcanisation of rubber, Goodyear 1841
Mechanisation of iron wood-screws, Colbert 1845
Sewing machine, Howe 1846
Rivet hole punching machine (developed for construction of the Menai Bridge) 1847
Lathe developments in USA, including turret lathe for armaments
Reaper put into production, McCormick 1848
Industrial vitreous enamelling process, C.H. Paris

Energy
Incandescent lamp, Groves 1840
Atmospheric gaslight burner 1840

Transport
90% of world's merchant shipping built of wood
50% of world's shipping built in USA and Canada
Steel cable suspension bridge, Roebling 1840
Thames Tunnel opened, M.I. Brunel 1843
First iron, screw-propelled, ocean-going steamship, *Great Britain*, I.K. Brunel 1843
Standard railway gauge introduced in Britain 1846

Communications
Calotype photography, Fox Talbot 1840
Micro-photography, Donne 1840
Penny Post commenced in England 1840
Paper positives in photography, Fox Talbot 1841

Architecture and Design
Corrugated iron roof 1840
Swedish Society of Industrial Design 1845
The Albert Dock, Liverpool, Jesse Hartley 1845
Palm House Kew Gardens, Burton and Turner 1847
Architectural Association founded, London 1848
Newcastle Central station, Dobson 1850

Society
Report analysing catastrophic sanitary conditions in England, Chadwick 1842
Chloroform anæsthetics, Simpson 1847
Communist Manifesto, Marx & Engels 1848
Denmark's first constitution (Grundlov) 1848
First public health act in Britain 1848
Cholera epidemics in London due to inadequate sewage system 1849 & 1853

Names
Richard Turner 1798-1881
M.G.B. Bindesbøll 1800-1856
Decimus Burton 1800-1881
W.H. Fox Talbot 1800-1877
Joseph Whitworth 1803-1887 *
Cyrus McCormick 1809-1884

Standardisation of the screw
The standardisation of size, thread and form of screws, bolts and nuts was fundamental to industrial development. The nail and the wood screw are 'self tapping' and require less precision than the machine screw where we are dealing with a system of male and female parts which depend on co-ordination of design in order to be effective. Production was at this time beginning to have a precision which both permitted and demanded standardisation and interchangeability of parts in a product within the *factory*. A part so universal in use as the screw obviously had to become standardised *nationally* – and preferably internationally. The inventor and machine tool manufacturer, Joseph Whitworth, saw this as an industrial responsibility and in 1841, after careful research into the prevailing confusion of different screw designs, he proposed a new standard specification of sizes and threads. This ultimately became the BSW (British Standard Whitworth) thread. Although it is based on the English inch it is still the dominating standard in many parts of the world. The two other international standards are Seller's (1868) in USA, and the metric thread (1898) of Europe which, with the general adoption of the metric system, is becoming increasingly important. Although standardisation of the screw seems to have happened surprisingly late in history, it is one of the earliest examples of industrial standardisation. The specification of the Whitworth standard is quoted below. See Taps & Dies and Hand Press pp 78-79

The photograph shows a set of taps and dies made by Whitworth & Co.
Photo: Museum of Science and Industry in Manchester Trust.
The drawing, showing the reverse side of the same screw stock, and the text below are taken from the Official Catalogue of the Great Exhibition 1851:

'Patent screwing apparatus, including the patent guide screw-stock and dies, working taps, master taps, for cutting up the dies, hobs, for cutting screw tools, and case-hardened tap wrenches. The dies of the screw stock are cut by a large master tap, and their action is the same as explained in the bolt-screwing machine. The diameter of the working taps are made to standard guages; the angle of the thread in all cases is 55 degrees rounded off top and bottom to two thirds of a complete angular thread; small fractional pitches are avoided, and the principle of uniformity in pitch, form of thread and diameter is rigidly adhered to'.

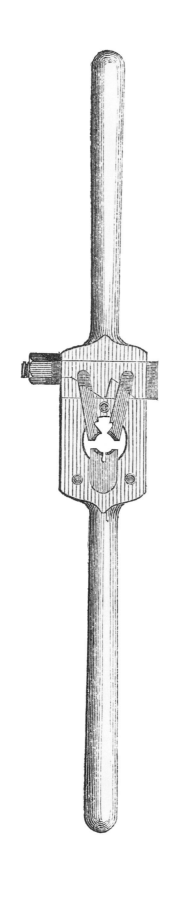

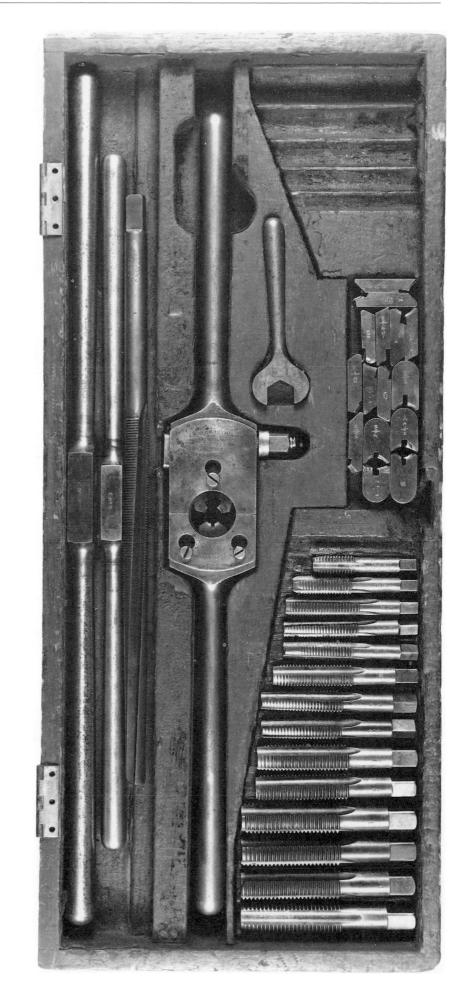

Entries marked with an asterisk are discussed in greater detail

Manufacturing

Sewing machine no.1, Singer 1851
Steel frame for umbrella invented, Fox 1852
Mass production of watches, Denison, Howard and
 Curtis 1853
Safety lock, Yale 1855
Factory for mass production of steam bent beech chairs,
 Thonet 1856 *
Bessemer converter, Bessemer 1856
Open hearth steel producing furnace, Siemens 1856
Circular kiln for firing ceramics, Hoffman 1858
Electroplating of iron and steel with non-corrosive metals
 commences
Automatic screw making machine patented, USA
Interchangeable production adopted in Britain
Brick making machine

Energy

Vortex turbine, Francis & Thomson 1852 (pp 56-57)
Atmospheric gas burner, Bunsen 1855
Theory of self-excitation of electric generator, Hjort
First (intentional) oil drill in Pennsylvania 1859
Electrical storage cell, Planté 1859

Transport

Mechanical ship's log, Siemens 1853
Great Eastern ship launched (19,000 t, with watertight
 compartments) I.K. Brunel 1858
Iron ships hull designed to calculated strain
Systematic oceanography and navigation

Communications

Telegraph cable laid under English Channel 1851
Electro-magnetic clock, Shepherd 1851
Multiple telegraph on single wire, Gintl 1853
Steinway & Sons founded, New York 1853
Colour photography, Zenker 1856
London press bureau founded, J. Reuter 1858

Architecture and Design

International trade exhibition, London 1851
Crystal Palace exhibition building, Paxton 1851 *
Paddington Station, I.K. Brunel & D. Wyatt 1854
International trade exhibition, Paris 1855
Gardner's Iron Building, Glasgow 1856
London's main drainage plan starts 1858
The Red House (W. Morris), Philip Webb 1859
Model factory town Saltaire, England, Titus Salt

Society

Census first showed more people living in towns than in
 country, England 1851
Crimean War 1853-1856
Science Museum, London, opened 1853
10 hour working day becomes normal in England, 12–14
 hours in rest of Europe
The Origin of Species, Charles Darwin 1859

Names

Michael Thonet 1796-1871 *
Joseph Paxton 1801-1865 *
Titus Salt 1803-1876
Isambard Kingdom Brunel 1806-1859
Henry Cole 1808-82 (planner of the Great Exhibition)*
Henry Bessemer 1813-1898
The brothers Siemens:
 E.W. 1816-1892, W. 1823-1883, F. 1826-1904,
 K. 1829-1906

The Crystal Palace

Whichever way one looks at it, Joseph Paxton's building which housed the Great Exhibition was a much more significant design and manufacturing achievement than most of the exhibits inside. Apart from being the largest building ever erected at the time, it was new in concept both structurally and technically and a remarkable contribution to the history of prefabricated building. Reading Digby Wyatt's detailed technical description of the building in the Exhibition Catalogue fills one with admiration. Referring to the 'airy lightness of the whole structure, and its immense dimensions' and the visitor's possible doubts as to its safety, he writes 'the whole of that which appears to him so complicated, is but the repetition of a few simple elements...'. As can be seen from the illustrations taken from the catalogue this is a column and beam construction of cast and wrought iron, timber and glass. The building is planned on an 8 ft module. Almost the entire structure is carried out with three column types, and their connectors, and three beam (or truss) types. The great vaulted roof of the 'transept' was added to accommodate a number of large elm trees which were growing on the site. Here the ribs and purlins are of timber. Paxton had previously made a special study of greenhouse roof construction so both the vault and the flat roofs are glazed with his patent 'ridge and furrow' system. The roof drainage design is an important part of the story but here we can only mention that all columns also served as downpipes from the roof. The quite new and untried building techniques motivated the invention of a number of machines for use both on and off the site, notably moulding machines for making the wooden guttering and glazing bars. The building measured 561 by 124 metres and took only just over six months to erect. *Punch* christened it 'The Crystal Palace'.

The illustrations are taken from volume I of the official catalogue of the Great Exhibition.

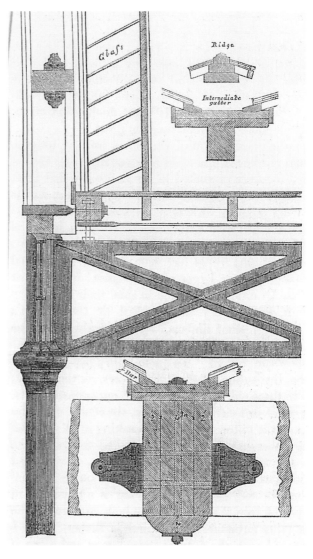

The main ribs of the transept vault consisted of a sandwich of three thicknesses of wooden segments sawn to the 11m radius and bolted together side by side with staggered butt joints, shown in the lower section.

View from the South entrance just after completion. The large elm trees are omitted. Note the 'ridge and furrow' construction of Paxton's special roof glazing. Height to top of vault 32 metres, width 22 metres. An interior colour scheme evolved by the architect Owen Jones exploited the long views of thin structural elements by painting them in shades of blue, red and yellow so that, according to the viewer's relationship to them, they blended variously giving an ever changing effect.

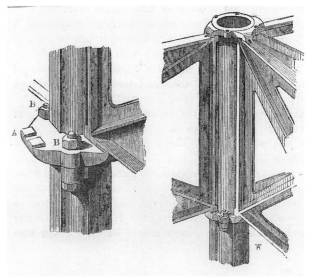

The cast iron column-truss connection.

Manufacturing

Whitworth's standard screw thread adopted in Britain
Regenerative tank furnace for glass founding, Siemens 1860
Automatic turret lathe, USA 1861
Slot drilling machine, Nasmyth 1862
Universal milling machine, USA 1862
Radial drilling machine, Whitworth 1862
Ball bearing design overcomes machining accuracy problems, A.L.Thirion 1862
Celluloid discovered, A. Parkes c.1862
Open hearth steel producing furnace, Martin 1865
Tungsten and vanadium steel, Mushet 1865
BASF founded 1865
Prohibition of lead in ceramic glazes in England
Hand binding harvester, McCormick
Mechanisation and mass production of clothes starts, primarily in USA

Energy

Oil: improvements in earth drilling techniques start
Refrigeration by absorption method, Carré 1860
Oil refining by 'cracking' for finer oils, USA 1862
Gas cooking and water heating introduced in England
Practical dynamo, Siemens 1866
Gas engine 1867
Dynamite, Nobel 1867

Transport

Asphalt paving 1860
London underground railway 1860-63
Petrol engine car, Marcus 1864 & 1875
Bicycle industry (crank and pedals on front wheel), Michaux 1867
Opening of Suez Canal 1869

Communications

Speed of light measured, Foucault 1862
Rotation press for continuous printing 1863
Motion picture, Ducos 1864
Theory of light and electricity, Clerk Maxwell 1864
Telegraph cable under Atlantic ocean opened 1866

Architecture and Design

International trade exhibition, London 1862
Clifton suspension bridge, I.K. Brunel 1864
Halles Centrales in Paris, Baltard 1854-1866
St. Pancras railway station, London, Barlow 1866
Reinforced concrete, Monier 1867
International trade exhibition, Paris 1867
Multi-storey workers' housing, London, Peabody 1860s and 1870s

Society

American Civil War 1861
1st Socialist International founded, London 1864
Das Kapital by Marx
Application of bacteriology in the food industry, Pasteur
Trades Union Congress founded (GB) 1860s

Names

Victor Baltard 1805-1874
Karl Marx 1818-1883
Louis Pasteur 1822-1895
Pierre Émile Martin 1824-1915
James Clerk Maxwell 1831-1879
Alfred Bernhard Nobel 1833-1896

Thonet Brothers

The pages immediately before and after The Crystal Palace deal with two quite different exhibitors at the Great Exhibition. They each represented the very best in industrial thinking and practice but were from opposite ends of the industrial spectrum. Whitworth, the fine mechanic and machine tool maker; Michael Thonet the artist chair-maker who threw himself – and his family – into the mass production of a completely new type of chair construction. Two of Thonet's products are analysed in Part II; here we are concerned with production. The steam-assisted bending of wood was traditional practice in boat building and some other crafts, including, to a limited extent, chair-making. Thonet's greatness lay in the application of steam bending to small and compound bends in chair parts as a fundamental element in chair design and production on a large scale. The detail which made this possible was the introduction, while bending, of a steel strap along the outsides of bends which relieved tension in the outer fibres and increased compression of the inner fibres of bends thus preventing fracture while bending. This process can be clearly seen in the top photograph. Some idea of the organisation, scale and design of bending equipment is given in the bottom photograph. Here chair front legs requiring only slight curves are steam bent without the use of tension straps. When air dried wood is heated right through to about 100°C, it takes on a completely different character. Its stiffness is transformed to resemble hard rubber and heating by immersion in steam is one of various methods of achieving this. After cooling and drying while still in the mould tool, it maintains its new shape. These bent parts are then simply joined with screws and bolts to become light, tough chairs. Michael Thonet had six sons each of whom specialised in the various aspects of the business. Five factories in all were established in S.E. Europe between 1856 and 1889 employing about 6000 workers. See Thonet settee and chair pp 140-143

Two processes in a Thonet factory of 1929.
Top: chairs could be supplied with various forms of bracket to increase their strength. Here wood, hot and flexible from the steam chamber is being formed in an iron mould with outer steel tension strap which can be seen transferring from one bend to the next.
Bottom: front legs with slight bends being formed.
Photos: Deutsches Museum.

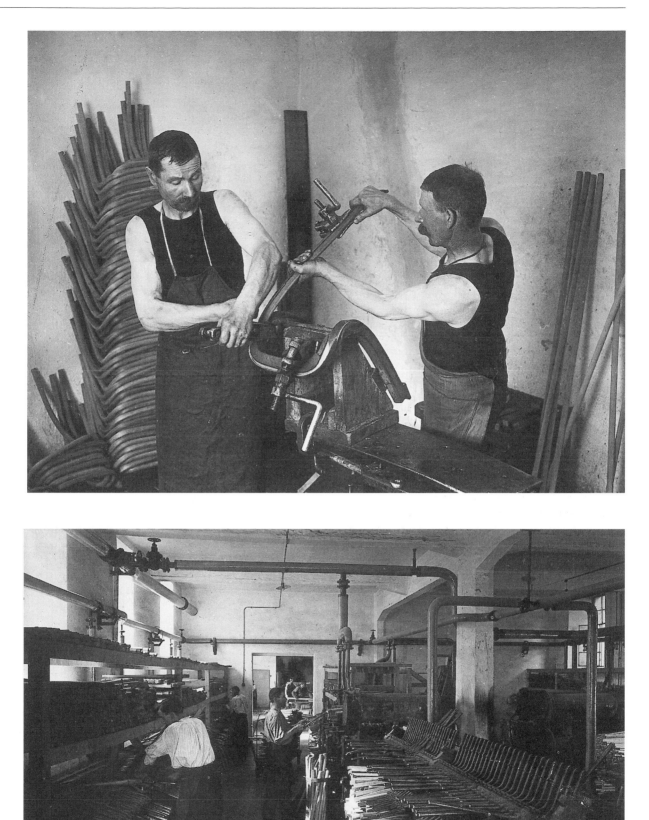

Entries marked with an asterisk are discussed in greater detail

Manufacturing
Electric steel furnace, Siemens 1870
Celluloid, Hyatt 1870
The cast-in-one earthenware w.c., Twyford 1870
Chair factory at Mount Lebanon, Shaker sect 1872
Automatic binder-harvester, 1875
Electric carpet sweeper, Bissell 1876
Steel tube for bicycle construction, Starley 1877
Chrome steel 1877 (chronologically: carbon steel,
    manganese steel, tungsten steel, chrome steel)
Automatic machine screw production, Spencer

Energy
Standard Oil Co.(standard refined quality) founded
    1870;'Refining' meant: 1.) 'dangerous petrol';
    2.) useful paraffin; 3.) residue, sold as fuel
Dynamo, Z.Gramme 1870
First Russian oil wells drilled at Baku 1873
Electricity begins to replace steam power for driving
    multiple belt shafts in factories
Ammonia compression refrigerator, Linde 1873
First electric cooking utensil, Lane-Fox 1874
Compressed air refrigeration, Coleman 1877
Carbon filament lamp, Swan 1878 *
Carbon filament lamp, Edison 1879 *

Transport
Model aeroplane, Penaud 1872
Tangentially spoked bicycle wheel, Starley 1874
Electric car, Siemens 1875
Standard time, USA railways 1875
First steel ship (as opposed to iron) 1877
Electric railway 1879

Communications
First typewriter made for sale, M. Hansen 1870-1875
First telephone patented, Bell 1876
Bell Telephone Company founded 1877
Microphone, Edison 1877
Phonograph, Edison 1877
First really functional typewriter, Remington 1878
World Post Association founded by 22 countries

Architecture and Design
The Arts and Crafts Movement at its height
Industrial design school in Copenhagen 1873 – becomes
    Denmark's Design School 1989
International trade exhibition, Vienna 1873
International trade exhibition, Philadelphia 1876
International trade exhibition, Paris 1878

Society
School reform: education for all, England 1870
The secret ballot introduced in England 1872
International Meter Convention 1875
Discovery of toxins 1876
600 boat passengers killed by water poisoning in river
    Thames 1878
London's main drainage system completed 1878
Migration from Europe to USA at its height

Industrialisation of tourism, Thomas Cook
Advances in municipal steam-powered water supply *

Names
Joseph Swan 1828-1914 *
C.M. Spencer 1833-1922
William Morris 1834-96
A.G. Bell 1847-1922
T.A. Edison 1847-1931 *

Ryhope Pumping Station
The basis of steam power is the vacuum which is created when steam is cooled by condensation in an enclosed chamber, and the pressure of steam when forced into an enclosed chamber. The steam engine is the arrangement which allows atmospheric pressure (resulting from the vacuum) and steam pressure to repeatedly and alternately operate on opposite sides of a piston in a cylinder, and converts the reciprocating action into energy. When this pumping station was built in 1868, the steam engine had been developing over 170 years; first to pump flood water out of mines, by Savery and then by Newcomen's beam engine; further improvements by Smeaton; improvements in design and power as prime mover for industry by Watt; the introduction of high pressure steam and adaption to locomotion by Trevithick; Stephensen's *Rocket;* and the development of the railways and the steamship. Ryhope was powered by a pair of very fine beam engines made by Hawthorn of Newcastle. They have high pressure cylinders of 70 cm bore by 152 cm stroke and low pressure cylinders of 114 cm bore by 244 cm stroke. Steam supply was at 35 lb per square inch with condenser vacuum 66 cm. The beam of each engine measures 412.5 cm between pump rod centres and weighs 22 tons. The flywheels each weigh 18 tons. The engines ran at 10 revolutions per minute and delivered 350 litres per stroke. In 1967 the engines were stopped for the last time after practically a century of continuous pumping. The waterworks is now open to the public. In his book *Beam Engines* T. E. Crowley writes, 'The larger examples are some of the most impressive of all man-made objects, and nobody who has ever seen one of them at work and has climbed two sets of stairs to watch a 40 ton beam rising and falling 12 feet in the air in almost perfect silence is ever likely to forget the experience'.

Ryhope Pumping Station.
Above: Plan and section of the building showing how the colossal engine, three storeys high, had to be built concurrently with the house. There are pump rods at each end of the beam, the one on the left pumped water from the main well which was piped to the staple well from which the right end pumped water into reservoirs.
Below: details of the beam engine.

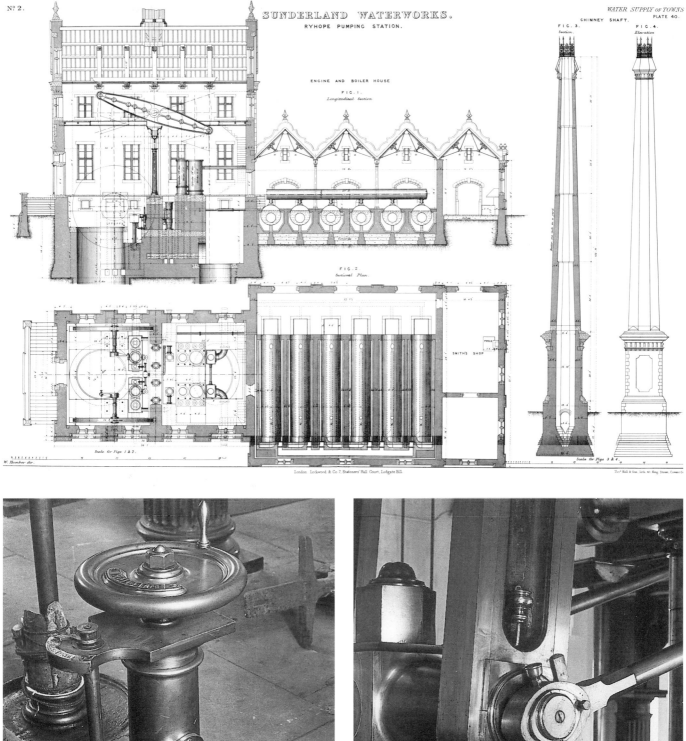

Nº 2.

SUNDERLAND WATERWORKS.
RYHOPE PUMPING STATION.

WATER SUPPLY OF TOWNS
PLATE 40.

CHIMNEY SHAFT.

ENGINE AND BOILER HOUSE

FIG. 1.
Longitudinal Section

FIG. 3.
Section

FIG. 4.
Elevation

FIG. 2.
Sectional Plan.

SMITH'S SHOP

Scale for Figs 1 & 2.

Scale for Figs 3 & 4.

W. Humber del.

London. Lockwood & Co 7, Stationers' Hall Court, Ludgate Hill.

Thos Kell & Son, Lith. 40 King Street, Covent Gn.

Entries marked with an asterisk are discussed in greater detail

Manufacturing
Grinding machine for steel balls for ball bearings,
    Fischer 1881
Borosilicate glass (heat, chemicals and electricity resistant), Schott 1881
Seamless steel tube patented, Mannesmann 1885
Aluminium production by electrolytic process, Hall and
    Héroult simultaneously 1886
Glass blowing machine 1886
Lever Brothers founded 1887
Barbed wire, USA

Energy
Steam turbine, De Laval 1882
First central power station, Edison 1882 *
Electric light bulbs with carbon filaments manufactured
    by Edison and Swan 1883 *
Gas lamp mantle, Welsbach 1885
Internal combustion engine invented, Daimler 1885
Deptford Power Station, Ferranti 1887
First electric flat iron 1888
First municipal electric power station, in Stockholm 1888

Transport
Electric elevator, Siemens 1880
Rover safety bicycle (prototype of the modern bicycle),
    Starley 1885 *
Petrol driven motor car, Benz 1886
Pneumatic rubber tyre, J.B.Dunlop 1888
Panama Canal built 1880-1889
First electric train and tram services

Communications
Half-tone block used in New York Daily Graphic 1880
Automatic table telephone, L.M.Ericsson 1882
Linotype, Mergenthaler 1884
International standard time 1885
Stockholm has more telephones than any other city 1885
Ballpoint pen patented
Hand camera 1886
Gramophone, Berliner 1887
Existence of radio waves demonstrated, Hertz 1888

Architecture and Design
Completion of Brooklyn Bridge, 1883
Steel frame skyscraper, Chicago 1884
Nordic Industrial, Agricultural and Art Exhibition,
    Copenhagen 1888
Galerie des Machines, Paris, Dutert & Contamin 1889
Eiffel Tower, Eiffel 1889

Society
International Patents Convention 1883
Aseptic surgery, Bergmann 1886
London County Council (L.C.C.) founded 1888

Names
Gustave Eiffel 1832-1923
Gottlieb Daimler 1834-1900
Heinrich Hertz 1857-1894

The electric light bulb
We have chosen the light bulb to represent the whole complex of the introduction of electricity. Initially it was the improvement of artificial light which was seen to be electricity's prime function. It is difficult to imagine the reception this incredible light source must have received. It was steady, noiseless, cooler, did not smell and did not exhaust the air in the room. The rigamarole of daily lamp lighting and maintenance was suddenly unnecessary; the light was brighter and to get it you simply pressed a switch. It took the ingenuity of glass blowers, engineers and scientists a period of about 40 years to achieve bulb designs which could be produced rationally and which were reasonably robust and long lasting. Its invention and design involved the solution to a completely new set of problems. Some of these were insoluble until other techniques had been mastered. For example, the filament had to be heated to incandescence without melting. This limited the choice of suitable materials since they all combine with oxygen when heated. This could only be prevented by surrounding the filament with a vacuum, but an adequate vacuum pump did not exist until the 1870s. At the end of all the research in several countries with countless materials and compositions two inventors emerge: Joseph Swan in Britain and Thomas Alva Edison in the USA. They arrived independently at closely related filament materials: Swan, carbonised extruded cellulose and Edison, rather surprisingly, carbonised bamboo fibre. But their designs were very different, Swan's lacking the simple practicality of Edison's. This is specially noticeable at the contact where Edison used the screw thread and Swan a spring and hook arrangement. This perhaps explains one of the idiosyncrasies of industrial design history: America and the continent of Europe use the screw-in bulb to this day, whereas Britain adopted the so-called bayonet fitting. Edison's contribution in this field also included electrical generating, distribution, and wiring circuits for buildings. Further developments in light bulb design accompanied those in electrification generally, and by the beginning of the 20th century tungsten became the accepted filament material. But it was a harsh light source which was harshly used. In the 1920s the Dane, Poul Henningsen, started research into the quality of light and ways of improving this in the room by the design and placing of the light fitting. See Pendant Lamp, pp 110-111.

Edison's light bulb with filament of carbonised bamboo fibre, c. 1881. Actual length 14.2 cm.
Photo: S.R. Gnamm, Die Neue Sammlung, Munich.

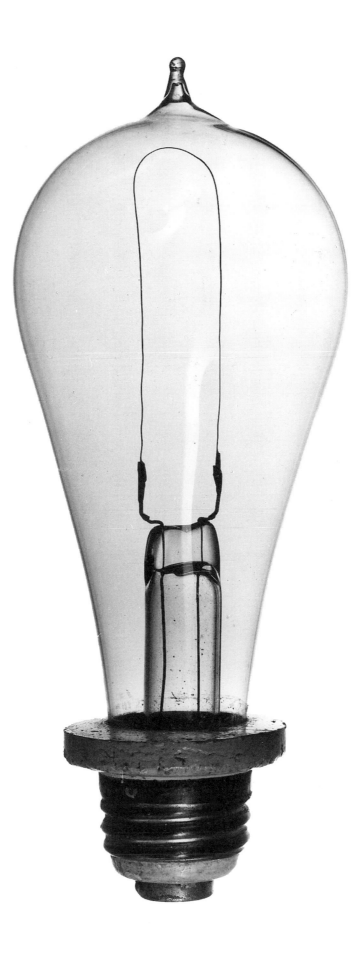

PART I. BACKGROUND AND INNOVATION 1890s

Entries marked with an asterisk are discussed in greater detail

Manufacturing

New grinding materials silicon carbide and aluminium oxide for the new high-speed steels

Gearbox for machine tools (speed selection and increase demanded by the new steels)

Automation in screw manufacture established by 1900

Multi-spindle lathes demanded by sewing machine production

Casein discovered in milk, leading to a form of plastic 1890s

Artificial silk from wood pulp, Cross, Bevan and Beadle 1892

Swiss army knife, Elsener 1897

Metric standard screw thread established 1898

Energy

Crystal Palace Electrical Exhibition, London 1891

Philips & Co. founded as electric bulb factory 1891

Thermos flask invented, Dewar 1892

Primus portable cooking stove, Lindqvist 1892

By-product coke oven, Hoffman 1893

Radioactivity, Becquerel 1896

Electric flat iron, AEG 1896

Osmium as electric lamp filament, Welsbach 1898

Radium, Curie 1898

Transport

Pneumatic tyres on bicycles 1890 *

Cunard liners climax of 19th century ship building: all steam, no sail

Kiel Canal opened 1895

Steam-driven aerodrome flight, Langley 1896

Diesel motor, R. Diesel 1898

Communications

Table telephone introduced, L.M. Ericsson 1892

Moving picture, Edison 1893

First international photographic exhibition, Hamburg 1893

'Phantoscope' first moving picture of modern kind, Jenkins 1894

Motion picture projector, Edison 1895

Radio telegraphy, Marconi 1896

Magnetic recording of sound, Poulsen 1898

Architecture and Design

First completely steel framed building, Chicago

Firth of Forth railway bridge, Baker 1890

International trade exhibition, Chicago 1893

L.C.C. multi-storey housing begins, London, 1895

The septic tank patented, Cameron 1896

Central School of Arts and Crafts, Lethaby first Principal, London, 1896

Workshop at Uccle, Van de Velde 1898

Society

The battle of Wounded Knee 1891

X-ray, Roentgen 1895

Nobel Trust founded 1895

The Garden City (book), Ebenezer Howard 1898

De Samvirkende Fagforbund (Danish TUC) 1898

Pain-reducing tablet aspirin comes into use 1899

Names:

Ebenezer Howard 1850-1928

William Richard Lethaby 1859-1931

Pierre Curie 1859-1906

Marie Currie 1867-1934

H.C. van de Velde 1863-1957

Valdemar Poulsen 1869-1942

The safety bicycle

The name 'safety' bicycle was used for some years after the so called 'penny farthing' lost its popularity and started to be replaced by the sort of construction we know today: two wheels of the same size, allowing the rider to sit lower and between them, and with geared-up chain drive to the rear wheel. The ingenious crossed, tangential wheel spokes were also a major strengthening improvement. The main industrial development which gave rise to this new concept was a satisfactory production of the steel tube, which with its high stiffness/weight ratio, was ideal for the bicycle frame. John Kemp Starley developed his Safety Rover machine with the help of this new material, first as shown in the drawing and in 1888 with a forward bend on the front wheel fork which greatly improved steering and balance. Some 200 more or less experimental pedal driven conveyances were produced mainly in France – which can be regarded as the home of the bicycle – and elsewhere during the two decades up to the end of the 1880s. Finally the pneumatic tyre (J.B. Dunlop) and the free wheel (Ernst Sachs) brought the bicycle substantially to its present layout by the end of the century. The intense activity in bicycle design at the end of the 19th century is well illustrated by the two designs shown. The photographed machine was probably made by Marriot & Cooper of London. The model appears in the M. & C. catalogue of 1889, but as the same type was made in France, Germany and Denmark there is a little uncertainty as to origin. Judging by the foot-rests on the front fork this model is still without free-wheel but has got pneumatic tyres and a brake. The bicycle led the way in its use of both the cog chain and the pneumatic tyre. It was the first mass-produced machine which directly contributed to the freedom of the individual and put personal transport within the means of the many. In all its unpolluting simplicity and in the way it honestly displays its structure and function, it is one of the longest lived and most useful representatives of the industrial functional tradition. As will be seen in the next article, early bicycle technology was to lead to design on an even higher plane.

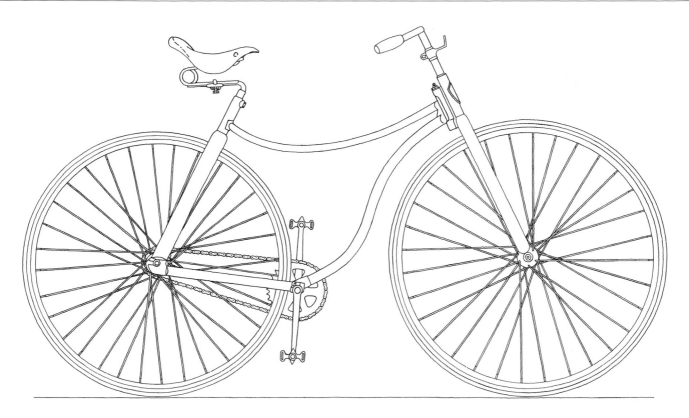

John Kemp Starley's Rover Safety bicycle of 1885.

The Humber cross-frame safety bicycle, c. 1895.

Entries marked with an asterisk are discussed in greater detail

Manufacturing

High-speed tool steel, Taylor and White 1900
Automatic glass bottle machine, Owens 1904
First truly synthetic plastic, phenol formaldehyde (PF), Baekeland 1907
Royal Dutch Shell founded 1907
Duralumin invented (alloy of aluminium, copper, magnesium, manganese), Wilm 1909
Time and motion study, F.W. Taylor, USA

Energy

Electric light bulb without vacuum, Nernst 1900
Russia world's largest oil producer (12 mil.t.) 1901
Thermos flask in production, Dewar 1904
Electric light bulb with tantalum filament, Von Bolton 1905
Upright vacuum cleaner patented, Spangler 1908

Transport

First zeppelin launched, F. von Zeppelin 1900
Metro underground railway opens, Paris 1900
Mass-produced motor car Oldsmobile, Olds 1901
Radial type aeroplane engine, Manly 1902 *
Ford Motor Company founded, Ford 1903
Oil burning steamship 1903
First man-lifting aeroplane, O. and W. Wright 1903 *
Flight in self-constructed biplane, Ellehammer 1906 *
General Motors founded 1908
Model T automobile, Ford 1908
Flight accross English Channel, L. Bleriot 1909 *

Communications

Transatlantic wireless telegraphy, Marconi 1901
Radio telephone 1903
Rotary duplicator, Gestetner 1903
Neon signs first displayed 1905
Television photograph, Korn 1907

Architecture and Design

International trade exhibition, Paris 1900
Wiener Werkstätte founded 1903-1932
Glasgow School of Art, Mackintosh 1897-1909
Copenhagen Town Hall, Nyrop 1902
Casa Batlló, Gaudi 1906
School of Arts and Crafts & Design in Weimar, H. Van de Velde 1906
Pennsylvania railway station, New York, McKim, Mead and White 1910
Deutcher Werkbund founded 1907-1934
Art Nouveau established as modern style
Paris Metro station entrances, Guimard

Society

The theory of modern psychoanalysis founded, Freud 1900
Death of Queen Victoria 1901
National Bureau of Standards, USA 1901
British Engineering Standards Association 1902, later British Standards Institute
Radiography, Marie Curie 1910

Names

Antonio Gaudi 1852-1926
Michael J. Owens 1859-1923
Leo Hendrik Baekeland 1863-1944
Henry Van de Velde 1863-1957
Hector Guimard 1867-1942
Wilbur Wright 1867-1912 *
Koloman Moser 1868-1918
Charles Rennie Mackintosh 1868-1928
Joseph Hoffmann 1870-1956
Jacob Christian Ellehammer 1871-1946 *
Orville Wright 1871-1948 *

The aeroplane

When contemplating the modern passenger aircraft with its superb technology and finish, it is difficult to believe that it all started, only a lifetime ago, as little more than an extension of kite flying, a hobby for the imaginative mechanic fascinated by aerodynamics – the wind provided lift, the motor should in principle be able to replace the kite string. The kite and the glider gave man the insight to understand that engine-propelled flight *must* be possible. He was egged on, a little at a time, by the occasional success in countless, often rather strange attempts at flight. In view of the subject discussed on the previous page, it is interesting that both of the world's first flyers were bicycle makers: the Wright brothers in USA made pedal bicycles and Ellehammer in Europe made motor bicycles. Ellehammer was a Danish inventor, educated as a clockmaker but with a very broad mechanical skill. His first aeroplane was entirely of his own design and making including the engine whose cylinders were arranged radially. It was this light, powerful radial motor which was Ellehammer's main contribution to aviation and the type widely used during the following years. His first flight, achieved after numerous accidents, repairs and adjustments, was from a circular concrete runway with a long rope holding the machine to a post in the centre of the circle. This was quite likely regarded by him as a sensible, reasonably safe stage in a series of experimental flights leading to the final free flight. But it was this 'assisted' flight which, at least in Denmark, is regarded as the first European aeroplane flight. It was followed a few days later by Santos Dumont in France. Ellehammer subsequently designed and made several successful machines including a monoplane, a seaplane and a sort of helicopter. In 1908 he won first prize in a flying competition near Kiel with a flight lasting one minute. He was the only competitor to leave the ground! The aeroplane is the first of the three great 20th century innovations. The two others, plastics and the computer, were to have a tremendous impact on its development.

The biplane with which Ellehammer flew for about a minute at a rally near Kiel in 1908.
Photo: Denmark's Tekniske Museum, Helsingør.

Comparing paleotechnic design characteristics with those of the neotechnic phase Lewis Mumford writes in 1934:
The gross over-sizing of standard dimensions, with an excessive factor of safety based on a judicious allowance for ignorance, is intolerable in the finer design of airplanes; and the calculations of the airplane engineer must in the end react back upon the design of bridges, cranes, steel buildings: in fact, such a reaction is already in evidence. Instead of bigness and heaviness being a happy distinction, these qualities are now recognized as handicaps; lightness and compactness are the emergent qualities of the neotechnic era.

Entries marked with an asterisk are discussed in greater detail

Manufacturing
Zip fastener invented 1913
'Pyrex' heat resisting glass, Corning Co. 1913
Cellulose Acetate plastic (CA), the first thermoplastic
    that could be injection moulded

Energy
Electric light bulb with tungsten filament, Coolridge
    1913
Horizontal, cylindrical vacuum cleaner, Electrolux 1915
Gudenaa hydroelectric power station, Denmark, complet-
    ed 1919, opened 1921 *
Splitting of the atom, Rutherford 1919
Basic construction of domestic electric appliances estab-
    lished
Electric smoothing iron

Transport
Loss of R.M.S. *Titanic* 1912
First Ford T from assembly line 1913 *
Traffic lights first installed 1914
Panama Canal opened 1914
Modern type construction of motor cycle
Citroën type A motor car 1919

Communications
First panchromatic film shown in USA 1913
I.B.M. founded 1912

Architecture and Design
Central railway station, Copenhagen, Wenck 1911
Design and Industries Association, London 1915
De Stijl movement, Holland 1917
Bauhaus founded in Weimar, Gropius 1919

Society
The existence of vitamins, Hopkins 1912
First World War 1914-18
The October Revolution 1917
League of Nations founded 1919/20

Names
James B.Francis 1815-1892 *
Peter Behrens 1869-1940
Walter Gropius 1883-1969

Hydroelectricity
Important developments in the generating of electricity by water power were made during the last half of the 19th and beginning of the 20th centuries. Electricity had the special capacity to be made in remote regions where there was a fall of water, or where a head of water could be created by damming, and then be conveyed in cables quickly, safely and conveniently over great distances to the consumer. The rapid rotation necessary to generate electricity motivated innovation in turbine design: where there was coal, the steam turbine was developed; where nature provided continuous water supply in quantity, it was the water turbine. The turbine is a simple motor whose power comes from the direct application of hydraulic force on a vaned wheel or runner. The turbine is normally submerged below the head of water which provides the pressure. Earlier turbines had simply formed wheels – comparable to the old water wheel though much smaller – and the wheel was not encased. As a result of hydrodynamic research in the 1850s, a great increase in efficiency was achieved by the development of the vortex wheel which later was enclosed in a helical flume casing. In this way the form of the vanes and the interior of the surrounding casing worked together to increase velocity and thus the output of the turbine. Research into hydrodynamics backed by advances in the fabrication and shaping of large complicated shapes in metal has consistently led to increased output with small-er heads of water. The Francis vortex shown here is one of the best examples of this.

The establishment of a hydroelectric plant differs radically from most other methods of generating in that here the prime mover is nothing more than the water and the turbines. The civil engineering works involved in getting the water to the turbines often forms by far the greatest part of the task. In the case of the Gudenaa project, where we found the turbine wheel pictured on the right, this involved the creation of Denmark's largest artificial lake, 13 km in length, behind an 800 m dam.

Now, 75 years later, the Gudenaa hydroelectric station still prides itself on being one of Denmark's largest, single sources of sustainable energy. At the same time its lake has been found to act as a pollution trap which is threatening the natural life of the whole length of the Gudenaa river. So here is a direct confrontation of environmental factors which clearly illustrates the complexity of the sort of ecological decisions which have to be made now, at the end of the 20th century. See Wind Power pp 68-69

The vortex wheel of the Francis turbine originally installed in 1919 at the Gudenaa hydro-electric power station in Denmark. This is one of a pair measuring 140 cm in diameter, one at each end of a 6 m long shaft. There are three such turbines, the shafts of which go straight through to the three generators. These have an annual output of 11 million KWh which is equivalent to the consumption of 3000 households. Siegfried Giedion wrote: 'The Francis turbine is still used to handling large volumes of water in low head installations. In the shaping of its blades and its whole construction, it is a plastic expression of the behaviour of smoothly flowing water'.

Entries marked with an asterisk are discussed in greater detail

Manufacturing

I.C.I. (Imperial Chemicals Industry) founded 1926
Polyamide plastic (Nylon), Carothers 1927
Colgate merges with Palmolive-Peet 1928 (two 19th century American companies)
Unilever founded 1929 (merger of British and Dutch 19th century companies)
Urea-formaldehyde plastics (UF)

Energy

The Aga solid fuel cooker, Gustaf Dalén

Transport

Diesel motor ship crosses Atlantic, Sweden 1925
Flight from New York to Paris, Lindburgh 1927
Last Model T Ford, no. 16.5 million leaves Manchester factory 1927 *
Flight accross the Pacific, Kingsford-Smith 1928
Bridge building programme in Switzerland, Maillart 1929-38

Communications

Radio broadcasting 1920
B.B.C. (British Broadcasting Company) 1922, became British Broadcasting Corporation in 1927
Electrically driven and amplified gramophones 1925
First talking motion picture 1927
Radio television 1927
First radio sets with earphones or loud speaker
Standard A-Format paper system, D.I.N.
Spring loaded pencil, Caran D'ache 1929
Advertising begins to dominate the environment

Architecture and Design

The Schröder House, Utrecht, Rietveld 1924
Furniture and interior design department founded by Klint at the academy of Fine Arts, Copenhagen, 1924
Communal design and housing programme under City Architect, Frankfurt 1925-30
Pavillon de l'Esprit Nouveau, Le Corbusier 1925
Bauhaus moves to Dessau. School building by Gropius 1925
Cabinet Makers' Guild furniture exhibitions, Copenhagen 1927-1966
The Copenhagen Exhibition 1929

Society

Insulin injections 1922
Fascist regime in Italy under Mussolini 1923-1943
General Strike in Great Britain 1926
Complete women's suffrage in Britain 1928
Penicillin discovered, Fleming 1928

Names

Henry Ford 1863-1947 *
Nils Gustaf Dalén 1869-1937
Robert Maillart 1872-1940
Knud V. Engelhart 1882-1931
Gerrit Rietveld 1884-1964
Kaare Klint 1888-1954

The motor car

When considering this phenomenon the famous Model 'T' Ford makes a good starting point. Here we have one of those milestones in the history of design and manufacturing where the product and the man behind it are of great significance. Henry Ford was mechanic, engineer and innovator. Brought up on a farm, his first achievements as a young man were the construction of a steam-powered tractor and a steam road vehicle, but he found that steam demanded a much too heavy and cumbersome construction. At this time in the 1880s the internal combustion engine had just been invented and oil was beginning to be refined. This, together with Ford's interest in finding a lighter, simpler solution, led to his first petrol-driven motor car. In his book *My Life and Work* he writes that he received little encouragement with his project; progressive people around him were convinced that the future lay in electricity rather than the internal combustion engine. Ford's commitment to the design of a transport vehicle was also set back by the popular reaction to the birth of the motor car; it was regarded as a sort of toy for racing. Speed was the most important thing, and this was sad, writes Ford, because it prevented people from making good cars.

The Ford Motor Company came into being with the help of a number of partners in 1903, and made various models which, like its competitors Packard and Olds, were largely handmade, expensive and based on assembly from sub-contracted parts. But this was never what Ford wanted and during these years he was designing the 'T' and planning its production. In 1906 Ford said, the 'greatest need today is a light, low-priced car with an up-to-date engine of ample horsepower, and built of the very best materials... It must be powerful enough for American roads and capable of carrying its passengers anywhere that a horse-drawn vehicle will go without the driver being afraid of ruining his car.' This was Ford's design concept. One of his main business principles was never to borrow money and to conserve profits for design and manufacturing development. His concept for mass production was power, accuracy, economy, system, continuity and speed – and good wages. His basic belief which permeated the whole Ford industry was that the motor car should be sold at the lowest possible price compatible with the right quality and the right design. He regarded product, manufacture and money as three interrelated, logical factors each of which has its own socially progressive rules.

The Model 'T' was the realisation of this manufacturing philosophy and behind the product lay countless innovations in the course of a production run which lasted some 20 years. The design and production of the Model 'T' can be divided into two periods which are related to production volume and the building of new factories. The first cars built between 1908 and about 1912 were slightly different in detail from those made under full assembly line principles from 1913 to 1927 as illustrated here. In all 16,500,000 Model 'T' cars were made, a figure which bears witness to the correctness of Henry Ford's production and sales policy: 'We cannot conceive' he wrote 'how to serve the customer unless we make for him something that as far as we can provide,

Model T Fords in production at the Trafford Park factory, Manchester, England, in 1914. This started by being an assembly factory only but by 1914 the wooden frame and sheet steel bodies were being made here. Production of Model Ts ceased in 1927. Photo: Ford Motor Company Limited, Brentwood, England.

will last forever.' So the 'T' became a constant and reliable 'institution': it was designed to be maintained and repaired by the owner; if your 'T' did finally wear out you could always buy another. Sales, however, fell abruptly at the end of the 1920s in competition with other manufacturers who were imbued with the fetish of the 'new model'.

Ford did more than initiate mass production, he did it in the spirit of the best of the early industrial craftsmen like Polhem, Maudslay and Whitworth. We have examined a Model T from 1922 and spoken with its owner who has driven many thousands of kilometres in it. All Henry Ford's principles as to quality of design, materials and manufacture are realised. Due to the nickel content of the steel used for the body and mudguards, they hardly rust; its friendly roominess contains two comfortable 'sofas' on which one sits like a lord; the whole generous, robust vehicle only weighs 700 kg. When looking at this simple, beautifully worked out motor car it is hard to believe that it represents the transition from horse drawn to motor driven transport. It also proved to be the beginning of an environmental misery.

Entries marked with an asterisk are discussed in greater detail

Manufacturing
Polymethyl methacrylate plastic (PMMA), or acrylic,
  Hill & Crawford 1933
Melamine-formaldehyde plastic (MF), Henkel 1935
Low density polyethylene plastic (polythene), ICI 1935
Polystyrene plastic 1937
Polyurethane plastics, Bayer 1937
Epoxy plastics
Hard and soft fibre board comes into general use

Energy
Domestic washing machines first appear in USA
Gas turbine invented 1930
Electric razor, Schick 1931
Fluorescent lamp invented 1934
The domestic refrigerator becomes popular in USA
Battersea Power Station, London, 1929-1955 *

Transport
Jet engine patented, Whittle 1930. Jet flight 1939
Parking meter invented 1932
London Passenger Transport Board founded 1933
Traction Avant motor car, Citroën 1934
Belisha beacons, L.H. Belisha 1935

Communications
Radar invented 1935

Public television service in Great Britain 1936
Bell table telephone 300, Dreyfuss 1937
Ballpoint pen first manufactured 1938

Architecture and Design
Stockholm Exhibition, chief architect Asplund 1930
Society of Industrial Artists founded, London 1930
Århus University, Fisker & Møller 1931 onwards
Tuberculosis Sanatorium at Painio, Aalto 1933
International Trade Exhibition, Paris 1937
New underground trains (Northern Line), London
  Transport's design team 1937
Standard unit kitchen furniture, Zwart 1938
Johnson Wax Building, Wright 1939

Society
The Third Reich 1933-1945
Bauhaus closes 1933
Second World War 1939-1945
Technocracy as political movement at its height 1932

Names
Frank Lloyd Wright 1867-1956
Frank Pick 1878-1941
Gunnar Asplund 1885-1940
Kay Fisker 1893-1965
C.F. Møller 1898-1990
Wilhelm Wagenfeld 1900-1994
Henry Dreyfuss 1903-1972

This panorama of metropolitan technology contains from the left: The backs of remarkably fine workers' housing from the mid-19th century in Peabody Avenue; Southern Railway cleaning sheds; Battersea Power Station (1929-1955); Southern Electric railway lines to Victoria Station (1908); the chimney of the Main Drainage Western (steam) Pumping Station (1875) – from palæotechnics to neotechnics at a glance.
Photo: Peter Moore/Stills Publications

Battersea Power Station

'As we have seen the great majority of technological developments were the result of empirical discoveries by practical men; indeed, it has been remarked already that until comparatively recently technology contributed more to science than science to technology. The electrical industry is exceptional in that its birth and development were the direct consequence of scientific research; moreover, the date of transition from experimental science to useful industry can be fairly accurately set. The key event was the practical demonstration of electromagnetic induction by Michael Faraday, announced to the Royal Society on 24 November 1831; within a very short time electromagnetic generators were being manufactured commercially.' (Derry and Williams in *A Short History of Technology*).

Generating plant has increased in size ever since and only a hundred years after the invention, power stations of the magnitude of Battersea are a part of every great city in the industrialised world. The building of Battersea stretched from 1929 to 1955 and from the start was the subject of controversy. The technocrats had recommended a building on a relatively central and prominent site, beside the Thames, of a dimension they were unable to handle architecturally. The architect, Giles Gilbert Scott, was called in to make cosmetic proposals which, in the event, were not entirely followed. It was closed down in 1983 and until recently stood derelict, like a gigantic gravestone to what we now can see was a misconcieved

technology. See Hydroelectricity pp 56-57, and Wind Generators pp 68-69.

One of the fascinations of cities is that every view that opens before one can be experienced as an historical statement. The view of London shown here, with all its curious beauty, is also a powerful judgement on the development of the industrialised city. We are only about two kilometers away from London's centre and yet there is not a soul to be seen and hardly anywhere they could be; the view contains nothing but technique – the services of living, but no life; the broad Thames is there but hidden, confined to its ditch on the nearside of the power station; the line of trees which show its meandering course is just visible in places. The air is misty with various forms of exhaust. Throughout the period covered by this book, London grew in size out of all proportion and here, in the electrical power station and the electric railway lines to the suburbs and dormitories, are two of the most important implements of its growth.

Entries marked with an asterisk are discussed in greater detail

Manufacturing
Aerosol spray invented 1941
Utility furniture programme, British Board of Trade
Teflon plastic (PTFE), Plunket (Du Pont) 1948
Glass fibre reinforced polyester plastic (GRP)
Electrolytic manufacture of tinplate
High frequency aided glueing in wood product industries

Energy
First nuclear reactor activated 1942
Ferguson tractor TE 20, Ferguson 1946
Microwave cooking developed 1948
New detergents appear on the market
The first large electricity-generating windmills *

Transport
2 CV motor car, Citroën 1946-1990
First supersonic air flight 1947
RT3 double decker bus, London Transport 1947 *

Communications
Computers – initial developments, USA 1942
Polaroid camera invented 1947
Transistor invented 1948
Long-playing records invented 1948

Architecture and Design
House of standard components, Eames 1949
Town Hall in Århus, Jacobsen & Møller 1942
Danish Co-op start furniture design department led by
    Børge Mogensen 1942
British Design Council founded 1944
Portex furniture programme for postwar rehousing, Hvidt
    & M. Nielsen 1944
'Britain Can Make It' exhibition, London 1946
Ergonomics in design starts (design of cab controls in
    army vehicles etc.)
Prefabricated building programmes for housing
Multi-storey housing, Bispeparken, Copenhagen

Society
Newspaper *Information*, Denmark, B. Outze 1943
Hiroshima and Nagasaki atomic bombed 1945
United Nations founded 1946
Mahatma Gandhi assassinated 1948
Antibiotics come into general use

Names
Harry George Ferguson 1884-1860
Charles Eames 1907-1978
Orla Mølgaard-Nielsen 1907-1993
Børge Mogensen 1914-1972
Peter Hvidt 1916-1988
Arne Jacobsen 1902-1971
Erik Møller 1909-

Public design and the RT3 bus
The quality and preservation of design standards in the public environment is one of the most important design fields there is, and it is tragic to see the political dismantling of a number of design conscious public services in later years, mainly in the transport sector. One of the earliest concerted public design projects on a large scale was that of the London Passenger Transport Board – commonly known as London Transport. The board was founded in 1933 to co-ordinate all public transport in London, principally underground trains and buses and their facilities. As part of this re-organisation good design was given a central role. Many professions were involved in the project not least important of whom were engineers, architects and designers, but in this case there was a single person who was none of these but without whom it probably would never have happened, the vice chairman of the Board, Frank Pick. He insisted upon, and inspired, an all embracing, recognisable standard of design: that is to say, a certain quality in the planning, form, graphic information, colour, materials and accommodation in buildings, street furniture and vehicles alike, which would lighten the stresses of Londoners' daily life and unify the vast, often chaotic, network which is London. The most conspicuous part of the project was the red double decker buses and by far the best designed of these was the RT3. The enormous replacement programme after the Second World War, called for a radical re-design of the prewar bus, based on quantity production, as opposed to the earlier batch production with more or less hand building techniques. The new construction and detailing were to a certain extent influenced by the aircraft design of the war and by the desire to improve the comfort and convenience of passengers, driver and conductor.

Since the London Transport initiative, similar design programmes have been completed in, for example, British Rail, the Milan underground train system, and the Danish State Railways (see pp 70-71).

Illustrations: The red double decker bus – London's 'trade mark'. The best of these was the 56 seat RT3 designed and launched 1945-1948. Structural members were mainly of light folded steel sections which were wood filled to resist buckling and to reduce weight. Aluminium was used for cladding and for internal fittings such as seat frames, handles and holding bars. Lightness with structural stiffness was one of the principal aims of the design. The interior was simple, direct and well detailed with specially designed upholstery textiles. The suspension technique was, however, unable to compensate entirely for London's poor road surfaces. This was specially noticeable when riding on the top deck.
Photo: London Transport Museum.
Drawing: Bus and Coach.

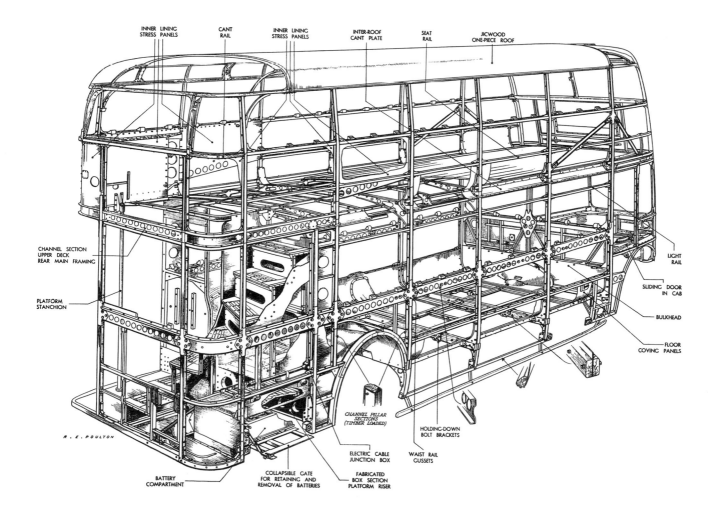

INNER LINING
STRESS PANELS

CANT
RAIL

INNER LINING
STRESS PANELS

INTER-ROOF
CANT PLATE

SEAT
RAIL

JICWOOD
ONE-PIECE ROOF

CHANNEL SECTION
UPPER DECK
REAR MAIN FRAMING

LIGHT
RAIL

PLATFORM
STANCHION

SLIDING DOOR
IN CAB

BULKHEAD

FLOOR
COVING PANELS

R. E. POULTON

*CHANNEL PILLAR
SECTIONS
(TIMBER LOADED)*

HOLDING-DOWN
BOLT BRACKETS

WAIST RAIL
GUSSETS

BATTERY
COMPARTMENT

COLLAPSIBLE GATE
FOR RETAINING AND
REMOVAL OF BATTERIES

ELECTRIC CABLE
JUNCTION BOX

FABRICATED
BOX SECTION
PLATFORM RISER

By the 1970s and 1980s, container traffic had become the normal method of transport of goods by sea. Shown here are 40 ft (12 m) containers and a container ship in Århus container dock, 1995. Another example of standardisation on an international scale.

Entries marked with an asterisk are discussed in greater detail

Manufacturing
Chipboard introduced in joinery, furniture etc.
High density polyethylene plastic, Ziegler
Polypropylene plastic, G. Natta 1954
Expanded polystyrene plastic, BASF

Energy
First nuclear power plant, USSR 1954
First large nuclear power station, at Calder Hall, England 1956
Nuclear power station accident at Kyshtym 1957
Nuclear power station accident at Windscale 1957
The ergonomically designed hand power tool, Zernell
Halogen lamp introduced for road vehicles

Transport
The container system of goods transport established *
Hovercraft invented 1955
Egypt nationalises Suez Canal 1956
British Railways modernisation design programme starts 1956
St. Lawrence Seaway opened 1959
Morris 'Mini' motor car, Issigonis
Diesel locomotives replace steam

Communications
Electric typewriter, Remington 1950
Videotape recorder invented 1956
Sputnik I, world's first orbiting satellite 1957
First transistor radios
Television becomes widely available
Colour television introduced
Standard A-format paper widely adopted in Europe

Architecture and Design
United Nations building in New York, Harrison 1947-1950
Festival of Britain 1951
International Council of Societies of Industrial Design founded 1957
Lousiana Museum of Modern Art, Denmark, founded by Kund W. Jensen, architects Bo & Wohlert 1958
Unité d'Habitation in Marseilles, Le Corbusier
Nôtre-Dame-du Haut, Ronchamp, Le Corbusier

Society
Credit card invented, USA, Ralph Schneider 1950
Contraceptive pill developed 1955
Treaty of Rome inaugurates EEC 1957

Names
Le Corbusier (Charles-Edouard Jeanneret) 1887-1965
Poul Henningsen 1894-1967
Alec Issigonis 1906-1988
Jørgen Bo 1919-1999
Vilhelm Wohlert 1920-

The container
One of the basic problems of all repetitive manufacture and one which increases with the quantities produced is the time and energy required to handle, stack, store and transport raw materials and products both in and outside the factory. The golden rule is to keep transfers from one location to another to a minimum. Within the factory, the means of handling depends on the type of production and product. Outside the factory the logical solution is the container in which goods can be transported in their entirety from point of origin to destination. This started in a piecemeal manner in the 1920s and 1930s and during the war was developed considerably. But for the method to become a worldwide system required co-ordination to an extent which affected industry and trade from the smallest detail of manufacturing and packaging to the planning and design of new mechanised harbour installations and container ships and vehicles. From the designer's point of view, one of the most fascinating things about the container 'revolution' is that the whole reorganisation centred around the container itself: it was the container dimensions which had to be decided and standardised internationally first, and then the remainder of the container complex could use this as its design module. This started at 244 x 244 x 602 cm long and was later increased to 1219 cm long. This reorganisation and planning of a new form of transport occurred, as it were, in silence, unknown to the general public (except for the inadequately explained dock strikes in the 1950s) but was, to our knowledge, the greatest example of practical international co-operation ever. One of the chief prerequisites for container traffic was highly controllable lifting power. In this respect the container marks the period of a pronounced increase in adaptable motive power, which has become normal within, for example, structural engineering, earthworks and agriculture. In the more domestic sphere the development of heavy lifting vehicles has enabled container collection of rubbish and of sorted glass, paper, plastic and metals for re-use and re-cycling.

Entries marked with an asterisk are discussed in greater detail

Manufacturing
Polycarbonate plastics 1960
Computer-aided manufacturing (CAM) begins *

Energy
Laser (Light Amplification by Stimulated Emission of
    Radiation) invented 1960
Industrialised, chemical-aided farming escalates and
    becomes the norm

Transport
Concorde, first supersonic passenger aircraft 1969
New design programme for road signs in Britain,
    Ministry of Transport and designer Jock Kinneir

Communications
Yuri Gagarin the first man in space, USSR 1961
Man lands on the moon, USA 1969
The first photocopiers for normal office use

Architecture and Design
Finnish Design Centre founded 1960
Palazetto dello Sport, Rome, Nervi 1960
Book, *The Measure of Man*, Dreyfuss 1961
Natural and artificial lighting laboratory started in School
    of Architecture, Copenhagen 1961
Berlin Philharmonic Hall, Scharoun 1963
Olympic sports buildings in Tokyo, Tange 1964
World Craft Council founded 1964
School of Architecture and Design founded in Århus,
    Denmark 1965
The last Cabinetmakers Guild Furniture Exhibition,
    Copenhagen 1966
Computer-aided design (CAD) begins *
Centre de Creation Industrielle founded in Paris 1969
    (moves to Centre Pompidou 1977)

Society
The Beatles pop group begins 1961
Berlin Wall erected 1961
DDT (first produced in 19th Century) proved to be
    cancer causing, Rachel Carson 1962

Names
Pier Luigi Nervi 1891-1974
Hans Scharoun 1893-1972
Henry Dreyfuss 1903-1972
Kenzo Tange 1913-
Some computer pioneers:
Vannevar Bush 1890-1974 *
Norbert Wiener 1894-1964 *
Howard Aiken 1900-1973 *
Walter H. Brattain 1902-1987 *
John Bardeen 1908-1991 *
William B. Sockley 1910-1989 *
Claud E. Shannon 1916- *

*The historical summary is based on chapter 8 of
'A History of the Machine' by Sigvard Strandh.*

The computer phenomenon
At this point in history, technology took a new turn. In
the context of this historical study of the way we have
shaped and made the implements of daily life, consideration of the computer requires us to take a very deep
breath. Here we are not dealing with just another branch
of technology, but one which has become all embracing,
and developed into a sort of master technology – not just
nuts and bolts, but something that was to have a dynamic
effect on people's daily existence on a worldwide scale.
Every walk of life was affected either directly or indirectly. In the attempt to mentally encompass the phenomenon
one looks back in history for equivalent examples with
which to compare it, but there are none. Electricity and
radio are the nearest we can get to it in magnitude. The
computer is by way of being the continuation of these.
Technology has always affected human thinking but only
through physical and material changes. Computer
technology is different. By taking over some of the
functions of the human brain, it radically changes our
relationship to the machine and to work, and it changes
our relationship with one another. It also changes the
communicational dimension of both our daily existence
and of the world. In the context of this book it is important not merely to accept the phenomenon but to consider
it critically. To start with, an abbreviated history.

The history of the computer
The abacus: the first step towards mechanising arithmeti-
    cal calculations.
1647 Calculating machine, Blaise Pascal.
1700s Various calculating machines attempted.
1830s The Difference Engine, Charles Babbage.
1840s The Analytical Engine, described by Babbage, was
    by way of being a theoretical programme for the mod-
    ern computer.
1870s-80s The Analog Computer, Lord Kelvin.
1920s Multi-calculating machine, Vannevar Bush.
1938 Electromagnetic relay calculating machine.
1944 The Automatic Sequence Controlled Calculator,
    Howard Aiken and IBM.
First generation computers:
1940s ENIAC - Electronic Numeral Integrator and
    Calculator. The first electronic computer. Very high
    energy consumption, taking about 200 sq m, low
    memory capacity, long access time.
1948 George Bowles' logical algebra of 1847 re-discov-
    ered and found to be ideal for computer technology.
1948 Cybernetics, or Control and Communication in the
    Animal and the Machine by Norbert Wiener. A general
    theory about all kinds of control including that of
    computers. The word cybernetics was invented by
    Wiener.
1948 The Mathematical Theory of Communication by
    Claude E. Shannon.
1948 The term 'bit' first adopted for the computer's
    basic yes/no, either/or unit of information.
1948 The transistor invented by Bardeen, Brattain and
    Shockley of Bell Telephone Laboratories. The word is
    made up from 'transfer' and 'resistor' which are
    basically its functions.
1951 UNIVAC (Universal Automatic Computer).

1950s Stagnation in development due to the difficulty of manufacturing transistors.

1956 'Leprechaun', first computer to use transistors, Bell Telephone Laboratories.

Second generation computers, comparable in size with other office furniture.

1958 Computer 7090, IBM, USA.

1959 Computer 803, British Elliot.

1959 Computer ER 56, SEL, Germany.

1960s The idea of the large, centrally placed, public computer, comparable with other local services. Idea short-lived due to the invention of the integrated circuit.

1964 First mass-produced integrated circuits (ICs), a fraction of the size of an existing circuit card.
Third generation computers, the beginning of mass computerisation.

1965 The first IC computers.

1967 Computer technology was now at a stage in which computers themselves enabled the development and mass production of the IC so that they became even smaller and more efficient.

1971 The microprocessor on a single 'chip'.

1980 The microprocessor strong enough to drive a small computer. The beginning of the spread of microcircuit technology into machine control in all fields, communications, and design work of all types.

From the list above we can see that it did not all happen quite as suddenly as we thought. As we have seen before in these pages the crises of war have repeatedly led to invention, and so it was with electronic control technology for the new weapons of the Second World War. Many of the pioneers involved in the first giant steps in computer development in the 1940s were, in a way, continuing their research from the war.

The year 1980 marked what could be called the end of the primary research period. Then followed the period of product development which increased the speed and the memory – the power – of the computer and decreased its size. At the same time it became more comprehensible and convenient for the user. In the mid-1980s as a result of an enormous design effort in the fields of perception and mental ergonomics came the 'what you see is what you get' principle in which the screen became the interface between machine and user, through imagery and text and the 'menu' and 'mouse' principle. The 'desk top' idea, too, sought to increase understanding by comparing what was happening on the screen with familiar office functions. The flat screen together with the minimising of electronic technique made possible the portable computer – which is writing these words with an ease, convenience and quietness one could only dream of when hacking away at its mechanical predecessor. The quality of image improved in grain and tone, and colour was introduced which made the computer into a better and better tool.

All this time a sort of snowball effect was taking place. The computer itself was speeding up its own development, manufacture and sale, added to which more and more uses were being found for it not only within data processing and recording, but in quite other fields such as servo steering and control of all types of machinery, increasing their performance manifoldly. Heavy capital products got heavier and more powerful; many consumer products became lighter and simpler. Known techniques and their machinery were thrown out (and their operatives with them) and substituted with computer steered equipment. Newspaper production was, naturally enough, the one we heard most about.

In the early 1990s the personal computer (PC) arrived and the snowball rolled even faster. Now we were beginning to look at a staggering commercial success and with it the waste of obsolescence on an unprecedented scale. Now a large proportion of the industrialised human race *had* to have a PC – both at work and at home. At the time of writing we are still at the stage in which the computer is regarded as much as a dilemma as an aid and yet we are accepting it strangely uncritically – almost as though it were a natural phenomenon. It is still a mixed blessing, and this is largely because it has not yet been properly integrated into society. We have not yet learned how to *design* computer technology or how to use it. It often acts as a substitute for proficiency. This is particularly noticeable in the poor quality of service from organisations which are there to serve, such as banks, insurance companies and public authorities.

The effect the computer is having on people is particularly noticeable to those who are concerned with design of all sorts, including architecture. It is a singularly good tool but it makes a distance between people and their creativity; the realities of life – real work, real environment, real materials – get out of sight and the screen world of the computer takes their place. One of the design fields which is suffering most obviously from the computer is graphics and layout. Here again it is making the impossible possible, but never before has one been confronted by so much confused, uncommunicative printed matter. As we have seen, computer technology now permeates all, from the writing and printing of this book to the wind generator and the IC3 train which are discussed on the following pages.

PART I. BACKGROUND TECHNOLOGY AND INNOVATION 1970s

Entries marked with an asterisk are discussed in greater detail

Manufacturing
Milan chemical plant disaster, 1976

Energy
Aswan Dam opened 1970
Coal miners strike over whole of UK 1972
Non-leaded petroleum introduced in USA 1975
Solar energy house, Philips 1975
Three Mile Island nuclear power station radiation leak, USA 1979
Windmill generators linked to national grid introduced in Denmark 1979 *

Communications
Compact disc comes on the market 1979
Photocopiers come into general use
Mini-computers introduced

Architecture and Design
Finlandia Hall, Helsinki, Aalto 1971
Olympic stadium, Munich, Frei 1972
Sydney Opera House, Utzon 1960-1973
Norwegian Design Council founded 1974
Book: *Det Siddende Menneske* (on the ergonomics of seating) Mandal 1974
Centre Pompidou, Paris, architects Piano, Rogers & Franchini 1977
Danish Design Council founded 1977
Legislation and design for improving the environmental integration of the handicapped

Society
Charter 77, Prague 1977
First test tube baby born, England 1978

Names
Alvar Aalto 1898-1976
Jørn Utzon 1918-
Otto Frei 1925-
Richard Rogers 1933-

Windmill generators
From where I am sitting writing, on a hill-top in Denmark, I can see through the window eight white windmills, similar to the ones shown here, spread over the countryside. They are generating electricity for their owners via the national grid. This is nothing new for the Danes. Generating windmills, often owned by farmers' co-operatives, were quite common in the early days of electricity at the beginning of the century. They were yet another example of the way new technology was quickly brought to the service of agriculture. Up to the oil crisis of 1973 Denmark's main power was almost entirely based on imported oil. Short-term alternatives were adopted such as the use of coal and North Sea gas. But long-term initiatives were also taken to harness sustainable sources such as sea waves, sun and wind. Research received government backing and in the case of wind turbines legislation was passed at the end of the 1970s to subsidise their installation and connection to the national grid. At this point the few small windmill engineering works developed into a major and technically very advanced industry. At the end of the 1990s there are 6 factories employing 9000 people (including sub-contractors). The industry has manufactured and installed 6000 windmills in Denmark and exported 14,170 all over the world, mainly to the United States.

The design, power and size of windmills has developed through the 1980s and 1990s from being comparatively low powered, with small turbines and a rather stumpy appearance, to being high, powerful and elegant. By 1997 they were being made in about 12 variations of power and height from 55kW at 30 m, to 1500kW at 80 m. A great deal of research has gone into turbine design which has led to increased size and output. The overall design of the nacelles differs according to the manufacturer. Some are nice and simple, others are less successful. But their contents and size are surprising when seen on the factory floor for the first time. They are big enough for a man to walk around inside. This is necessary for maintenance, access being up the inside of the column. They contain shafting through the main bearing from the turbine hub to a large, heavy gearbox. A flexible coupling from this (with a fail-safe hydraulic disc brake) turns the generator at the rear of the nacelle. In the examples shown, the generator is liquid cooled from the prominent air intake and radiator on top. At the main nacelle bearing at the top of the column are the yaw motors to keep the turbine facing up into the wind. Without computer technology the modern windmill generator as we know it would not be possible. Monitoring and all operations such as yawing, breaking, start and stop are done automatically and supervised by remote control.

Columns, nacelles and turbines are usually coated with off-white finishes. This combined with their elegance and simplicity makes them noticeable, but pleasantly unobtrusive in the landscape; in fact, in slightly misty conditions they become almost invisible. In this connection their correct siting and grouping are of great importance. The two main factors here, which are sometimes in conflict with each other, are the achievement of optimal wind force and regard for the aesthetic quality and type of landscape.

The importance of wind power to the world's environment cannot be overstressed. By 1997 installed wind power capacity in Denmark passed the 1000 megawatt mark, enough to produce 7% of the national electricity consumption. This has already reduced the emission of CO_2, one of the highest per head of population, by 3%.

Three 600kW windmills, with 45m hub height, made by NEG Micon. Situated in East Jutland.

Entries marked with an asterisk are discussed in greater detail

Manufacturing

Union Carbide factory disaster, Bhopal 1984

Health hazard due to asbestos established

Wood-bending technique by axial compression revived and mechanised, Danish Institute of Technology and Compwood a/s 1990

Computer numerical controlled machines in industry (CNC)

Energy

Low voltage halogen lighting for buildings introduced

Energy saving, compact fluorescent lamps introduced 1981

Chernobyl nuclear disaster, USSR 1986

Public awareness of the phenomenon 'greenhouse effect' 1988

Super tanker Exxon Valdez grounds in Alaska causing oil disaster 1989

8000 windmill generators exported from Denmark 1983-1989 *

3000 windmill generators in Denmark 1990 *

Transport

Advanced passenger train IC 3 by Danish State Railways (DSB)1988*

Zeebrugge car ferry disaster 1988

Tunnel connecting France and England opened 1996

Rail and road connection across Great Belt, Denmark 1998

Communications

Electronic typewriter, IBM 1981

Computers, improved and reduced in size, come into general use worldwide

Architecture and Design

Annual furniture design exhibitions (Snedkernes Efterårsudstillinger) founded, Copenhagen 1981

Danish State Railways (DSB) design programme evident throughout the country. Architect Nielsen*

Copenhagen Royal Theatre extension project, Fehn

Society

'Solidarity' movement in Poland starts 1981

Ethiopian famine 1984

Vienna Convention on protection of the ozone layer 1985

Aids epidemic begins

Brundtland Report, Strategy for a Sustainable Development 1987

Montreal Protocol on the reduction of CFC gases 1987 and 1990

Berlin Wall falls 1989

Names

Jens Nielsen 1937-1992

Sverre Fehn 1924-

The IC3 train

Since the 1970s the Danish State Railway, DSB, has steadily improved its standard of service and design. It was important that public transport did not continue to sink into oblivion in the face of the increasing popularity of the motor car, and that the real qualities of rail travel such as safety, economy, and less pollution be revived. In spite of media criticism the programme has been an enormous success. A small group of DSB executives, designers, engineers and manufacturers have put creative planning, design and technology at the service of a whole population and thereby made the railways popular again. The programme started with more or less cosmetic improvements to the existing rolling stock: greater comfort through good detailing and pleasant textiles in passenger compartments, a telling corporate image achieved with simple means, functional graphics and delightful posters; then came the stations. Buildings were restored and modernised or renewed throughout the country and platform signing, lighting, seating and other important details were given a strong, recognisable design treatment which gave passengers the feeling of being in the care of a national organisation. The latest achievement in this modernisation programme is the IC3 train. Design work started in the early 1980s and in 1984 the first full-size mock-up was built on the factory floor. Some ten years later the IC3 was in service throughout the country. This train represents a complete re-think as to the design requirements for a passenger train for Denmark, with its unevenly distributed 5 million inhabitants, living in regions which are separated by water. These conditions require light, fast train sets, each containing their own engines, which can be quickly coupled and de-coupled. In this way several destinations can be served without passengers having to change. The IC3 'feels' convincing. This simple, gloss white train with its crisply detailed coach work and generous, smoothly operating red entrance doors, glides in and out of stations swiftly and silently. On boarding you find yourself in a delightful and generously proportioned lobby which simply and naturally guides you to the right, or to the left, according to your seat number. DSB's chief architect who was largely responsible for the conception and realisation of the IC3 writes: 'The overall requirements of the interior of the IC3 are enough space and accessibility, visual clarity, flexibility and ease of maintainance. Over and above this, is the physical experience of travelling which is governed by the running properties and acoustics, and the comfort on board.'

Next on the programme came the electric version.

P.S. The manufacturer, Scandia of Randers, was founded in 1861 by three English railway constructors to supply wagons for the railways they built – some of Denmark's earliest.

Above: The IC3 is a three-car train set with automatic coupling which can be quickly assembled to form longer trains. The body shells are of extruded aluminium. A train set weighs only 90 tons. Motive power is provided by four standard lorry engines placed under the floor. An aggregate of 1600 HP, accelerating to top speed gives a 40% reduction in energy. This efficiency enables a cut in travelling time by 20% giving a corresponding rise in numbers of passengers.

Below: The key to the IC3 is at the ends of the train. Here are two conflicting functions: the driver's cab, and a comfortable passage when train sets are coupled. This conflict is brilliantly resolved by hinging the cab so that it swings 90° inwards thus opening the passage. The passage is sealed from weather and noise when two train sets are coupled and their inflated rubber frames are pressed together. Wind tunnel tests have shown the rubber ends to be aerodynamic.

PART II

METAL PRODUCTS

In this and the following parts of the book we take a close look at individual products in order to learn something about design and manufacture from them. Our descriptions of the way they are manufactured are sometimes brief, partly because of the difficulty of establishing and describing the manufacture of historical artefacts, but also in recognition of the fact that the best way for the designer to understand making and materials is to experience them. We cannot know it all, anywhere near, but we can acquire an understanding which will enable us to design well. It is this that has been our aim.

So this introduction will go through a few of the general properties of metals and their working, and manufacturing techniques. The fundamental things are often left out, so here we will look at some of these. Probably the most important thing for the designer to know about metals is that they have a closely held secret. They are malleable, that is to say shapeable, in the same way that clay is shapeable. What makes metal seem so formidable is its hardness and strength when handled *in human hands*. But if you change your conception of force and energy, and with the help of impact or pressure apply much greater *mechanical* force and expend much more *energy* on the metal, you can make it do whatever you want. The history of industrialisation is synomynous with making metals do what we want them to do, and improving their properties so that they do what we want them to do with other materials as well.

Some metals such as copper and tin are malleable in their cold state. Some metals can be mixed to become alloys which are malleable if heated to various degrees of glowing heat. Malleability also has to do with mass. If the mass is very small as, for example, with tinsmith work, all the necessary shaping of the material can be performed cold with small hand-turned rolling machines and the like (see Re-use cans, pp 102-103). If the mass of metal is great, heat and impact – that is to say forging – have to be employed. A later development here, however, is presses of such power (6000 tonnes and more) that large forgings of some 60 kg can be pressed, rather than forged.

Forging gives the greatest strength of all the physical treatments, both in stiffness and toughness. Casting is also a way of shaping metal but with quite different results: a normal casting is brittle. Casting, as opposed to forging, is performed with the metal in its molten state, being poured or injected into the mould, which is of the shape (in negative) of the object to be cast. Casting is extensively used, and in the case of castings made from a mould of compressed sand, this is clearly recognisable by its rough, slightly sand-like surface.

If we take the opposite extreme to casting and forging, and one which is of equal importance, we have rolled sheet metal. This is made in various thicknesses and widths according to purpose. The best known example of sheet metal is that from which motor car bodies and refrigerators are made. Sheet metal is the raw material for stamping, pressing, punching, embossing, and many other operations, many of which stiffen and strengthen the sheet. But all the components and products which are derived from sheet metal have a completely different character from castings and forgings. They are much tougher than castings: where castings are brittle, they are resilient; where castings are hard and stiff, sheet components are lighter and subject to vibration.

Metals react very differently to being machined and worked. They often change their molecular structure and therefore their character and workability. So they have to be annealed, tempered, and normalised by heating (and sometimes cooling) in different ways, to either regain their original character, or to become easier to work.

Metals and their alloys can be divided into two groups: those containing iron, and those which do not. This is a good grouping for the designer because those containing iron are the strongest, the most adaptable in production and generally the cheapest, but they rust – except for stainless steel,which is being used more and more. Those which do not contain iron, which are known as non-ferrous metals, have all the finer qualities which the irons lack, but they are not so strong. None of them rust.

Under the iron group are various qualities of cast iron and mild steel, which is purified iron with very low carbon content. A type of iron which has been very important throughout our story, but which is now obsolescent, is wrought iron. It has many of the qualities of steel, and more besides, but it gave way to steel which is cheaper and is adaptable as an alloy with other metals. Steel, again, is divided into two types: carbon steel and steel alloys. There are many qualities of carbon steel from 0.25% carbon, which is the most used of all metals, to 1.3% carbon steel which is used for sharpenable cutting tools.

The introduction of steel alloys in the mid-19th century gave industrialisation a big push forward because they increased the capability of machines themselves and their performance. There are a large number of steel alloys, designed for many purposes. Three of the best known are: nickel steel, with its good resistance to corrosion, impact, and metal fatigue; stainless steel, the great improver of hygiene in hospitals as well as the home; and high-speed steel (wolfram, chrome, vanadium and carbon) for machining many materials – including metals.

There are many non-ferrous metals. We will look briefly at copper, tin, aluminium and zinc, and their alloys. Copper is the only common metal which has a reddish colour. It is a good conductor of both electricity (in cables, wire, switch gear etc.) and heat (cooking utensils, heat exchangers, etc.). Its self-protecting green layer when exposed to the elements, makes it both a beautiful and effective covering material for roofs, facades etc. Its most important alloys are brass (mainly copper and zinc), and bronze (no zinc, but tin and other materials, according to what is required of it). Copper and its alloys are used extensively in plumbing. Tin is one of the earliest plating materials and gives its name to tin-plate which, in fact is thin, tinplated steel. Aluminium has become increasingly important because of its lightness and its resistance to corrosion. Aluminium is extremely soft and therefore has to be mixed with other metals (copper, zinc, nickel, chromium, manganese – to name a few) which enable it to be made in all forms from rolled sheet to castings. In this respect it compares with steel, but is not nearly so strong. Zinc is well-known as galvanised plating which gives very good protection to iron and steel under normal weather conditions. As zinc alloy it has a number of uses both in its capacity as a weather resister and as castings.

The working of metals from raw material to finished product

The manufacturing pattern of a metal product or component varies greatly according to material, shape and purpose, and the number and types of industries involved. Each product has its 'family tree', parts of which will resemble the diagram below. The diagram is meant as a guide to some of the main processes involved and their relationships. It does not show industrial structure, and it does not fully cover the much more complicated field of mechanical design. Some of the processes and materials are specified by the manufacturer of the final product, some are part of the constant production of 'raw materials' in standard form and in large quantities.

Many of the materials and processes named here are to be found in the products which follow. The history of the introduction of metals and their alloys is often indicated in the events lists in Part I.

17th century folding cutlery.
Photo: S.R.Gnamm, Die Neue Sammlung Munich.

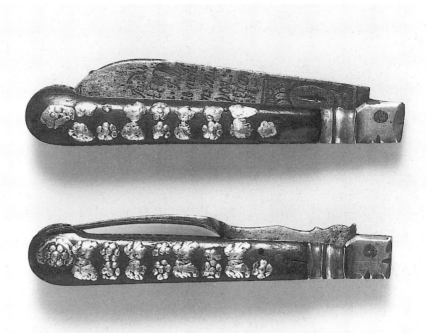

Mid-19th century travellers' cutlery with bone handles by Harrison Bros. & Howson, Sheffield.

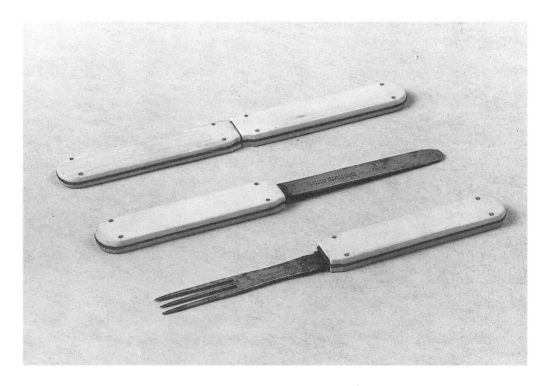

Subject: A fork (opposite) which was designed to be packed for travelling when inns expected everyone to bring their own cutlery. There is a knife to match and they will have fitted into a leather case.
Manufacturer: Unknown but English.
Designer: The silversmith.
Materials: Silver with steel tips spliced to the tines.
Dimensions: Length when screwed together 19 cm.

Location: City Museum, Sheffield, England.
Evaluation: The design of compact cutlery for travellers to bear on their person or in their luggage showed remarkable innovation and advanced technique, as here with the screw solution and the beautifully executed steel tipped tines from about 1725. With its strong styling, the fork seems to embody both the past and the future.

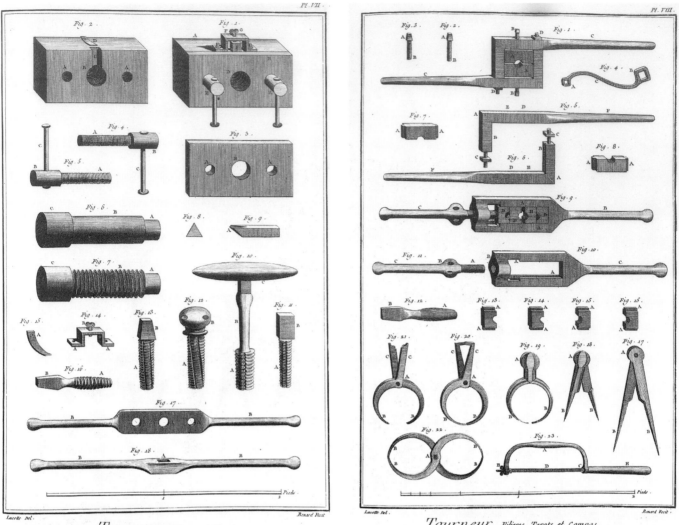

From L'Encyclopédie, ou Dictionnaire Raisonné des Sciences, des Arts et des Métiers, by D. Diderot.

The screw: One of the fundamental devices on which mechanisation is based is the screw. So in the study of the history of industrial design it makes good sense to consider an idea which has accompanied the whole of design development. It started over 2000 years ago as a powerful means of applying pressure and tension and, rotated in a pipe, for conveying certain materials such as water and corn. Much later it became a universal fastening device as well. The taps and dies shown above are for making internal and external screw threads and they bear a remarkable likeness to present day hand threading tools. The lefthand plate shows taps and dies (or 'screw boxes') for large diameter threading in wood for presses, vices and the like. The righthand plate shows metal threading tools for making screws and bolts. The metal screw was a prerequisite of industrialisation. It not only connected the parts of products and in so doing pressed them firmly together, but it enabled their separation again. Parts could be mass produced and assembled with screws and bolts to become complete products thus

showing the way to a basic principle of industrial production: interchangeability of component parts. This means that any component from any production batch can be assembled, without further fitting, to any other component from any other batch to complete a product as intended. This conception and the ability to put it into practice had its beginnings at the turn of the century. Curiously enough another 40 years were to pass before a similar accuracy control was put into effect for the screw. See Standardisation of the Screw, pp 42-43.

Hand press: Research into the origin and purpose of this hand press has been disappointing, but in spite of this we cannot resist including it because of its astoundingly timeless quality. Its function as a very powerful hand press is clearly and beautifully expressed by its form. This and the fact that it is described as a 'knee press' could be clues to its purpose: it has been suggested that it could be a demonstration model for the training of engineers.

Hand press presented to the Conservatoire National des Arts et métiers, Paris in 1866 by the Société d'Encouragement à l'Industrie Nationale. The machine dates from the late 18th century. Materials are brass, steel and wood. Height 40 cm. Photo: S.R. Gnamm, Die Neue Sammlung, Munich.

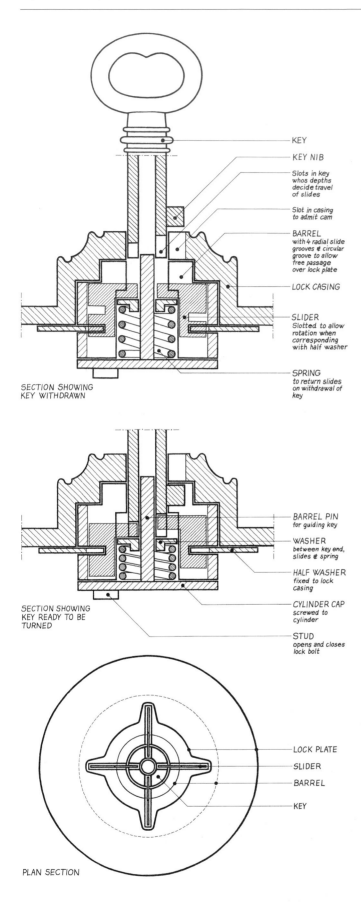

KEY

KEY NIB

Slots in key
whose depths
decide travel
of slides

Slot in casing
to admit cam

BARREL
with 4 radial slide
grooves & circular
groove to allow
free passage
over lock plate

LOCK CASING

SLIDER
Slotted to allow
rotation when
corresponding
with half washer

SPRING
to return slides
on withdrawal of
key

SECTION SHOWING
KEY WITHDRAWN

BARREL PIN
for guiding key

WASHER
between key end,
slides & spring

HALF WASHER
fixed to lock
casing

CYLINDER CAP
screwed to
cylinder

STUD
opens and closes
lock bolt

SECTION SHOWING
KEY READY TO BE
TURNED

LOCK PLATE

SLIDER

BARREL

KEY

PLAN SECTION

Measured drawing by Martin Bohøj (not to scale).

Subject: Lock mechanism for various applications such as doors in buildings, cupboard doors, box lids, drawers and padlocks.

Manufacturer: Joseph Bramah, London. The firm still exists, now under the name Bramah Security Equipment Limited.

Designer: Joseph Bramah patented the mechanism in 1784 and then again in 1798 when machines had been designed and made to manufacture the locks.

Materials: Mostly brass. Slides, spring and lock plate of steel. Key, iron or steel.

Manufacturing processes: Sawing, stamping, casting, planing, filing, slotting, bending, screw tapping and turning, spring winding, polishing.

Dimensions: The photos show a 9 in. door lock and a 3 in. chest lock. Cylinder casings 19 mm diameter.

Location: The door lock is in the Science Museum, London. The chest lock is in the author's collection.

References & notes: The text and illustration of Bramah's patent of 1784.
The Construction of Locks, A.C. Hobbs and Charles Tomlinson, London 1868.
The Machine Tool Collection, Science Museum, London, by K.R. Gilbert 1966.
Joseph Bramah, Ian McNeil, Newton Abbot 1968.
Eighteenth Century Inventions, K.T. Rowland, Newton Abbot 1974.

Evaluation: With the help of the measured drawing of the chest lock, which is slightly simplified for the sake of clarity, it can be seen that the key depresses four spring-loaded steel strips (sliders) which slide in deep grooves in the brass cylinder. Slots in the sliders are arranged to prevent the rotation of the cylinder, and hence the bolt from moving unless they are depressed to their different positions by the correct key. The compactness of the Bramah lock is achieved by its radial arrangement, the key enters at the centre of the mechanism which operates radially when the key is pushed in, against resistance from the spring, and turned. The key is small and devoid of pocket-wearing protrusions. The solidity and precision with which the lock and key function has to be tried to be appreciated. This is characteristic of many early industrial mechanisms which were often machine-made but hand-fitted and finished. The number of slots in the key corresponds to the number of sliders which varies between 4 and 12 according to the degree of security required. The number of combinations can amount to hundreds of millions.

The Bramah lock is historically unique. It was invented before the necessary machines existed to make it accurately and economically, so Bramah, with his foreman Henry Maudslay, constructed some of the very first machine tools to manufacture the locks. Over 200 years later the locks are still in production and are one of the most secure on the market. In 1980 the Company produced a very neat padlock in which the cylindrical casing follows, quite naturally, the form of the mechanism.

9 in. door lock, late 18th century. Photo: British Crown copyright, Science Museum, London.

3 in. chest lock and stricking plate late 19th century.

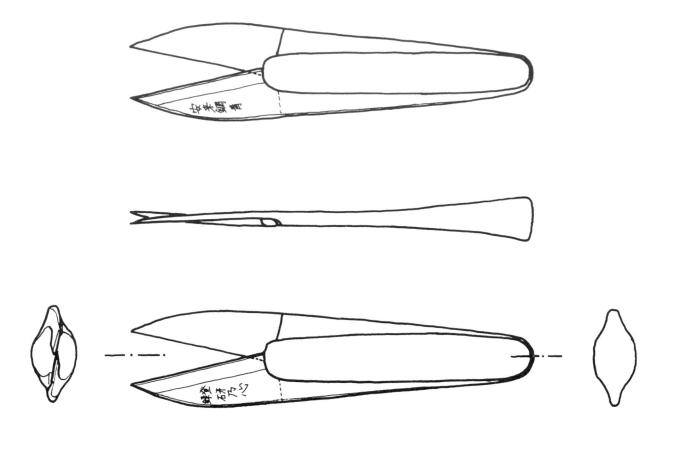

Measured drawing of the Tokyo type by Akis Nakagawa (full size).

Subject: The normal type of scissors for light work. Three regional variations are shown in the photo, from the left: Kyoto, Osaka, Tokyo.

Manufacturer: Fujii Metal Ltd., Ono, Japan.

Chronology: Origin probably ancient Egypt or Greece; via Silk Road to Japan before 1100; at the beginning of the 19th century concentrated production was started by several smithies in Ono; these examples are from the 1980s.

Design: Vernacular.

Manufacturing process: Traditional: made singly by forge-welding carbon steel to the cutting edges of the knives and welding to wrought iron bar, forging to shape, grinding, polishing, bending to U form. Modern: Knife parts A and B are stamped out of a composite strip of mild and carbon steel. These are welded one to each end of a mild steel bar. The bar is hot pressed to shape a rough ground C, and polished and blackened by a soda, phosphate process. Final grinding and polishing. Bending to U form.

Dimensions: Length 10.5 cm. In various sizes.

Location: Author's collection.

References & notes: *Brugsting Fra Japan,* Magnus Stephensen and Snorre Stephensen, Copenhagen 1969. *The Scissors Book,* Yuji Sano, 1987, Japan.

The History and Manufacture of Nigiri-Basami, an illustrated report by Akis Nakagawa, Osaka, 1989.

Evaluation: The nigiri-basami (grip scissors) is an ancient tool which in Japan preceded the X-formed scissors by some 100 years. It has developed with the sword and edge-tool technology in which steel is forged to one edge of an iron strip to form a composite blade. The steel provides the keen, sharpenable cutting edge and the iron provides a tough, resilient backing. This is a technique which was highly developed in Japan as far back as the 9th century and which was also employed in Europe. In relation to the U scissors design this is of particular interest because, while the technique has become less important due to the availability of cheap steel and its alloys, the U scissors have continued to employ it to provide handle, blades and spring in one. Akis Nakagawa writes: 'The product development of U scissors in Japan has been influenced by this technique. It is as though the technique has determined the beauty of form and function in the nigiri-basami.' The modern method of production was developed by Hiroshi Fujii in 1950. It maintains all the traditional qualities and enables a 50% increase in production. Unpractised Western fingers could easily and quickly cut the outline of the Japanese word for 'scissors' in paper (opposite).

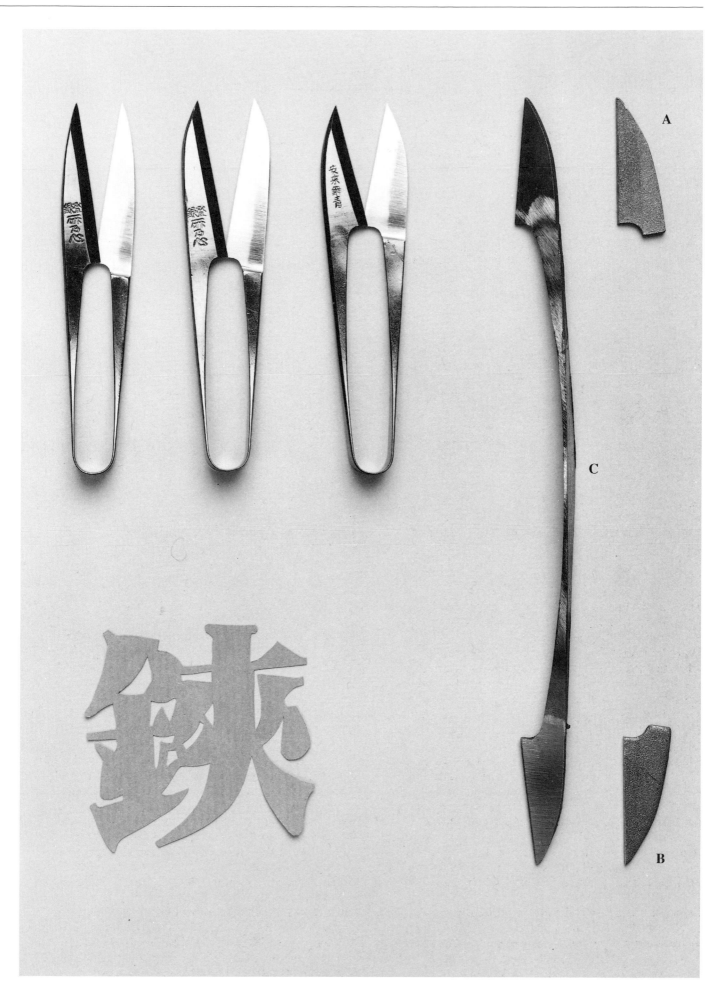

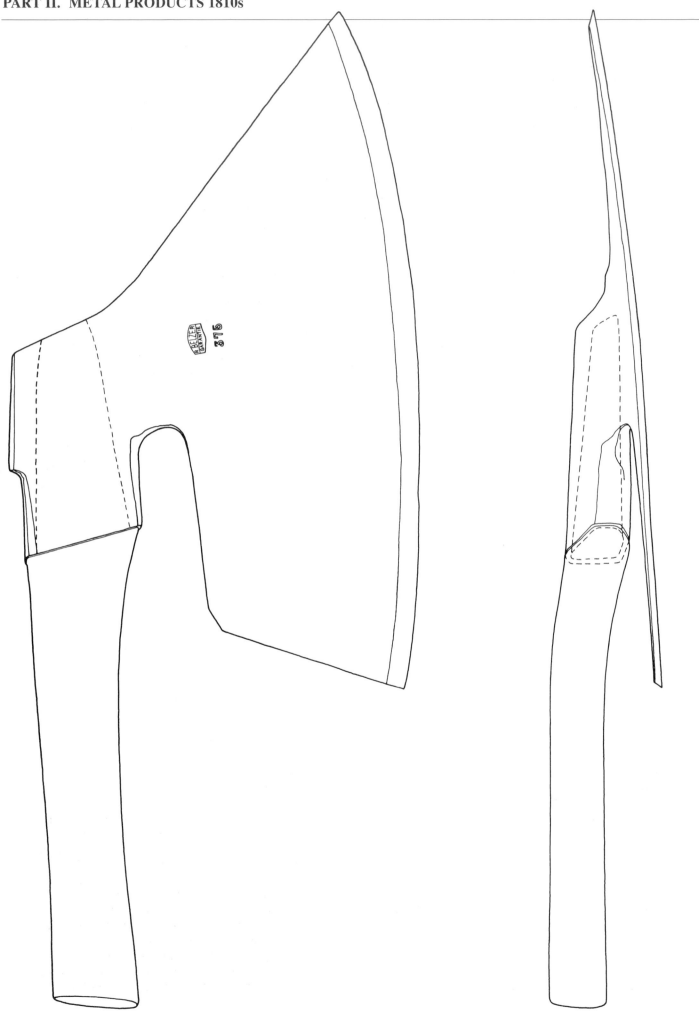

Subject: Carpenter's side axe, or broad axe, for converting round timber to rectangular sections for columns, beams, joists, rafters etc.

Manufacturer: R. Beltzer. Probably German.

Design: Traditional continental European pattern.

Materials: Forged steel blade, wooden handle.

Manufacturing processes: Forged, ground, polished, bevelled. The handle is made by the craftsman owner who also does the final sharpening.

Dimensions: Cutting edge 36.7 cm. Weight 2.5 kg.

Location: Private collection.

References & notes: *Dictionary of Tools used in the woodworking trades, 1700-1970,* by R.A. Salaman, London 1975.

The publications of Arnold & Walker, Needham Market, England.

A Museum of Early American Tools by Eric Sloane, New York, 1973.

Gammelt Værktøj by Børge Dahl, Roskilde, 1974.

Left: measured drawing by Nis Øllgaard, scale 1:2

Evaluation: This is a particularly fine representative of the great variety of side axe patterns to be found in Europe and America and developed through hundreds of years. The actual date of manufacture of this example is not known, but similar patterns were in regular use throughout the 18th and 19th centuries and, on the continent of Europe, right up to the middle of the 20th century. Their function was taken over by the large bandsaws and circular saws of the sawmill which did not, however, give the same smooth finish. Now they are used almost solely for restoration work. The tool is used vertically in short downward strokes, the user walking slowly backwards to the side of the work as he hews. The curved, chisel edge cutting diagonally across the grain gave the characteristic rippled or wavy surface to beams, columns and other structural members in half timbered and old wooden buildings and ships. The handle is held with both hands and is formed so that it curves a little away from the work to give space for the knuckles. The whole tool is a superb example of what could be called the ergonomics of experience, as can be clearly seen in the drawing. Industrialisation has so changed the method by which we convert timber that it is hard for us now to realise what an important and common tool this was.

Subject: Imperial Standard capacity measures. The cylindrical containers are for dry measurement, the conical containers are for liquid measurement.
Producer: The British government.
Chronology: Both the conical measures below and the bushel measures in the drawing are from 1824. The dry measures to the right are from 1870.
Materials: Brass. Ebony handle-grips on the dry measures.
Manufacturing processes: The cylindrical dry measures are cast and turned. The conical liquid measures are also cast and turned but stand in lifting trays of formed sheet material.
Dimensions: See drawing, in mm. The measures are extremely heavy in themselves; the half bushel one, for example, is hard to lift.

Imperial measures of capacity:

4 gills	=	1 pint
2 pints	=	1 quart
4 quarts	=	1 gallon
2 gallons	=	1 peck
4 pecks	=	1 bushel
8 bushels	=	1 quarter
36 bushels	=	1 chaldron

Location: The cylindrical measures in the photograph are at the Conservatoire National des Arts et Métiers, Paris. The conical measures are at the Science Museum, London. The drawn measures, which are just as imposing in their design, are in the Tide Mill in Woodbridge, England.
Evaluation: By the beginning of the 1800s the units of measurement in England were approximately at their present value. In 1824 the Imperial Standard gallon of 277.42 cubic in. was established, and with it the whole series of capacity measurements represented here. The conflict between the policy of free enterprise, and the increase in trade due to escalating industrialisation, required the establishment of state-imposed standards of measurement which were unquestionable, 'unadjustable' and respected. Hence the design, quality and sheer weight of these measures which express authority in every detail.

Liquid measures: quart, gallon and pint. Photo: British Crown Copyright Science Museum, London.

Dry measures: gallon, peck, pint, half pint, gill. Photo: S.R. Gnamm Die Neue Sammlung, Munich.

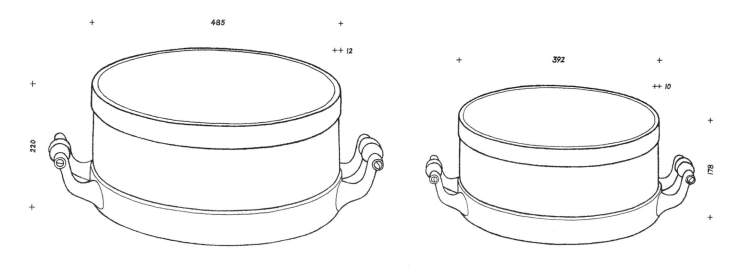

Bushel   *Half bushel*

Subject: Door latch and drawer handles.
Manufacturer: Keep & Hinkley, Birmingham, England. From their catalogue of 1845.
Design: Probably the manufacturer, though much copying prevailed at the time.
Materials: Brass (alloy of copper and zinc).
Manufacturing processes: The designs are based on contemporary fashion and on the skills and practices of the brazier: sand-casting, riveting, grinding, boring, filing, fine grinding, buffing, polishing. All hand and treadle work.
Dimensions: The illustrations are here slightly reduced from those in full size in the catalogue.
Location: The catalogue is in the Local Studies Department of the Birmingham Central Library.
References & notes: *The Victoria and Albert Museum's Collection of Metal-work Pattern Books* by Nicholas Goodison.
Evaluation: In 1816 there were 85 braziers and brass founders in Birmingham alone. They comprised a key

industry, supplying a large range of domestic equipment. Their products were normally of good quality and well-formed and, along with a number of industries of the period, the catalogues in which they were presented were remarkable. They were primarily for the use of wholesalers and travellers; note the prices per dozen. Most articles were drawn full-size in a specially developed etching technique in which one sometimes ignored the laws of perspective for the sake of a clearer description of every detail. The brazier was one of the craftsmen who laid the foundation to industrialisation. With his foundry and sand-casting technique, he was able to produce 'standard' products in so many different sizes that it was almost as good as getting things specially made. It gave architects and craftsmen much greater choice and design freedom than is the case today. These fittings are typical Birmingham products and were exported all over the world. In the 1950s these brass fittings were still revered and used by some of the famous cabinetmakers of Copenhagen.

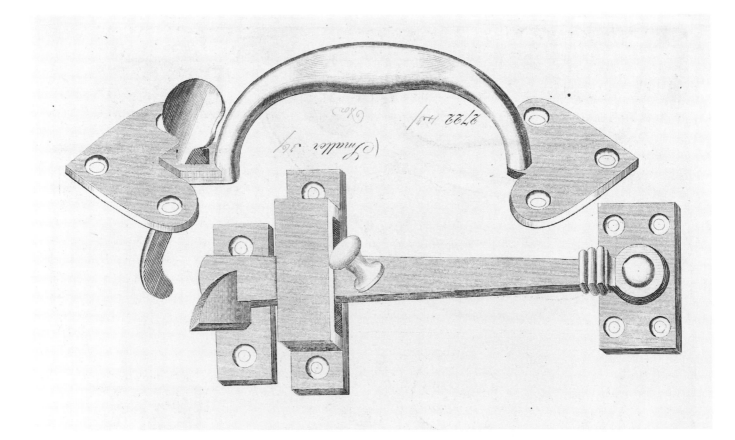

The illustrations are from Keep & Hinckley's catalogue of 1845. Apart from problems with perspective, illustrations were occasionally printed upside down, here corrected in the fall door latch above. Opposite: the catalogue contains many more handle sizes and types for drawers, chests, cases, doors etc.

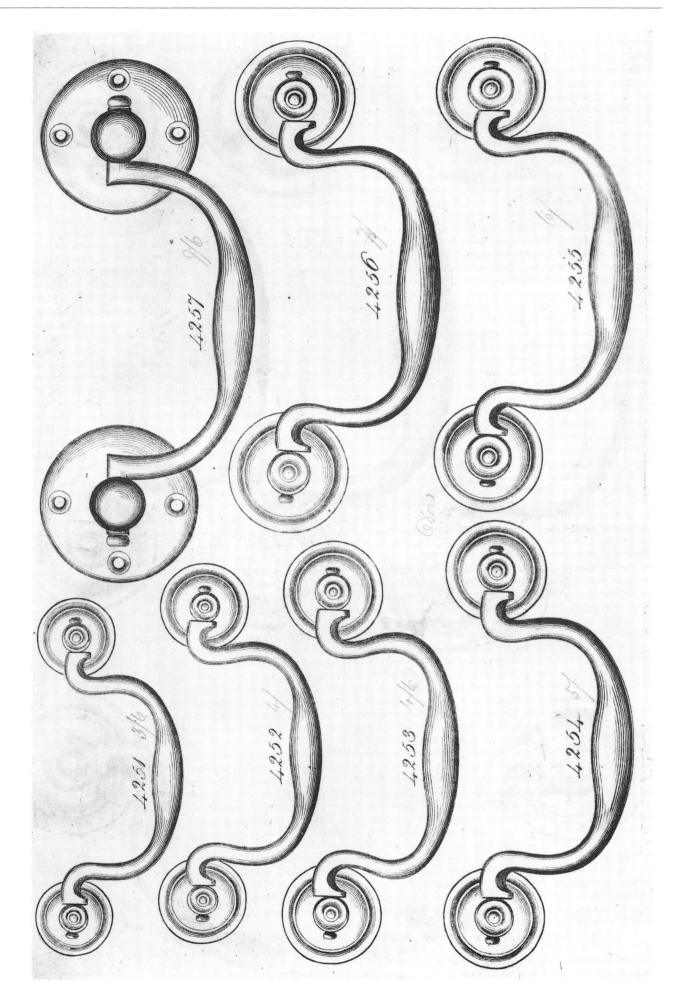

The 'fine' exterior ...

Subject: 8-day clock of the so-called 'Ogee' pattern, made from the 1820s to 1914. The name comes from the ogee section of the case frame.
Manufacturer: Waterbury Clock Company, Waterbury, Conneticut.
Design: Anonymous.
Materials: Movement frame and wheels of brass, steel axles. Dial of enamelled zinc. Case of coniferous wood, partially veneered.
Manufacturing processes: The movement is stamped, punched and bent from rolled brass sheet. The case is made with cheap joinery techniques, put together entirely with pins and glue with a minimum of material. The whole case is stained and the veneered areas polished.
Dimensions: Case made in several sizes this example being the most common. Width 39 cm, height 65.5 cm, depth 10.5 cm.
Location: The Open Air Museum, Den Gamle By (The Old Town), Aarhus, Denmark.
Refences & notes: *From the American System to Mass*

Production 1800-1932, by David A. Hounshell, Baltimore, 1984.
Evaluation: All through history time-telling has been a part of technical advance. The clock was a prerequisite of industrialisation. It not only told the time but synchronised people's activities and enabled them to organise. The development of the clock has demanded better alloys, tools and machines which have been quick to spread to other industries. It was a clockmaker who invented crucible steel for his springs; the invention of the industrial spinning machine was largely the work of a clockmaker; a clockmaker invented the planing machine. The American clock's claim to fame is due to it being the earliest large-scale, commercially mass-produced article, first, amazingly enough, with wooden movements and later, when the clockmakers had secured sufficiently good quality, rolled brass sheet, with movements of primarily stamped brass. This meant that the fine-toothed wheels of the clock movement instead of being hand-turned and filed from brass castings could be produced in

... and the crude interior, holding the stamped-out movement.

a fraction of the time. By 1845 the clockmakers of Conneticut State alone were producing nearly a million clocks a year. In 1850 the largest company sold 280,000 clocks and was exporting them all over the world. The construction and jointing of the case is the poorest imaginable, but the general conception is fairly sensible. The simple rectangular shape comes from the necessity to accommodate two weights, one for the power supply to the clockwork and the other for the chime. These go up and down, one in each side of the case, with their strings to the movement going over pulleys at the top. The outside, visible part of this product is one of the few we have chosen for which we have little respect. It is here to mark the beginning of commercial mass production and to represent the millions of inferior products which have been produced ever since.

However the movement, although on quite a different level from the clockmaker's art, is something of an achievement. Unfortunately, very little evidence of this very competitive industry remains but the machinery which made the clockwork must have been fairly sophisticated, requiring great accuracy and good tool making. The advent of the so called 'American System' is always discussed in relation to the American clock industry, and possibly has something to do with that special American ability to produce inexpensive, roughly made, light machinery which is none the less robust and reliable.

The clock is representative of the group of products in which the important part is an inner mechanism which is protected and concealed by an outer casing. In the American clock of the later 1800s this relationship followed the well-known lines of commercialism: the case was exploited as a marketing device, which by continual re-design created obsolescence and the ability to withstand competition, and increased sales. The parts that mattered – the works and the dial – remained more or less the same. A design contradiction in one and the same product.

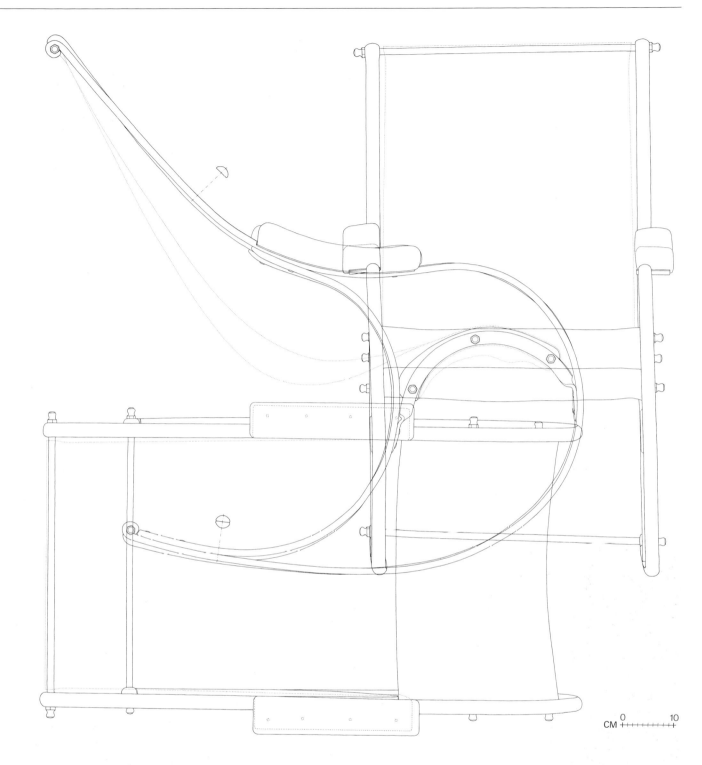

CM 0 ++++++++++ 10

Subject: Rocking or reclining chair.
Manufacturer: Uncertain, but R.W. Winfield & Co. of Birmingham, England, showed a similar design in brass at the Great Exhibition of 1851.
Designer: Probably R.W. Winfield and Dr Calvert.
Materials: Frame of wrought iron with domed nuts of brass. 'Hammock' of canvas covered in velvet.
Manufacturing processes: Blacksmith's work: heating, bending in jigs, forging, riveting. Upholstery work. Bolted assembly.
Dimensions: Width 57 cm, height 107 cm, depth 108 cm, seat height 49 cm. Weight 21 kg.
Location: Photographed chair: Clausholm Manor, Denmark. Drawn chair: Author's collection.

Evaluation: The very high seat and the somewhat help-less position one is forced into when seated could be explained by the chair's therapeutic connections. A Dr Calvert is associated with the design of the chair which he recommended for 'ladies and invalids' and described as a 'digestive chair'. The conception and design of the chair is remarkably clear and simple, and it must be one of the earliest examples of so-called 'knock-down' furniture: unbolt the five cross rods and the chair can be packed or stored flat. Its great weight makes it a difficult chair to handle and very hard on floors.

Measured drawing: Nis Øllgaard and Martin Bohøj.

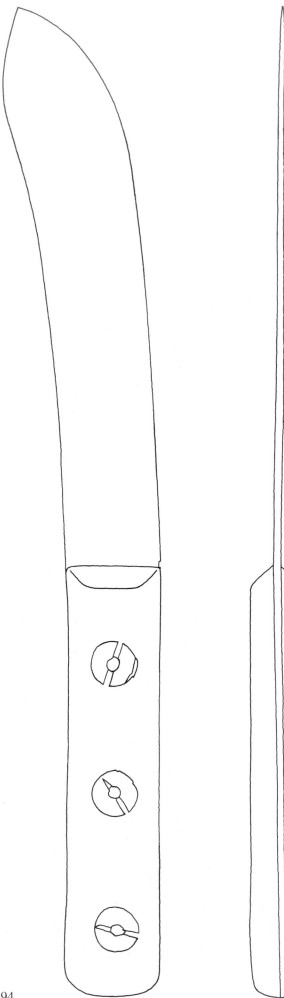

Subject: Butcher's knife, general purpose scissors.
Manufacturers: Knife: Francis Needham, Sheffield. This maker was in business between 1900 and 1935 so this example is later than the 1860s but a good representative of the vernacular.
Scissors: E. Hunter, Sheffield, 1851.
Designs: Traditional.
Materials: Knife: Carbon steel blade, ebony scales, brass bolts.
Scissors: cast carbon steel.
Manufacturing processes: Knife: Drop hammer forged from steel bar, hardened, tempered, ground. Handle: scales sawn to length from battens, drilled (also through haft) and counter sunk, assembled, ground flush, and rounded. Blade and handle emeried then polished. Blade sharpened.
Scissors: Cast, ground coarse then fine, buffed, polished, plated, edge ground (sharpened) all as separate blades. 'Put together' so that the blades shear evenly along the whole of their length with screw which is finally riveted. All work requiring great personal skill.
Dimensions: Knife length 26.5 cm, blade 15 cm.
Scissors: Total length 17.5 cm, blades 7.5 cm.
Location: Knife: Author's collection.
Scissors: City Museum, Sheffield.
References & notes: If the scissors are nickel plated, as they appear to be, they would have been one of the very earliest nickel-plated products.
The Story of Cutlery by J.B. Himsworth, London, 1953.
Related products: Domestic Cutlery pp 114-115.
Evaluation: A book should be written about the history of the functional tradition in Sheffield. Here was a town of workshops, dedicated to the making of tools and cutlery, which grew into an industrial city whose core remained the smiths, the grinders, the cutlers, the 'scissor put togetherers' – the small specialised makers of the implements for the cook, the farmer, the carpenter, the cabinet maker, the tailor and the home. Here are two typical examples of the functional tradition from Sheffield's heyday. They are proof of the fact that really good product design and making remain relevant indefinitely. The scissors are nearly 150 years old and they could have been made yesterday.They have to be used to be believed. A similar butcher's knife is still made. It is the most rational and strongest possible way of making a knife. It is a delight to use and hold. The grey steel, the black ebony and the yellow brass rivets express most satisfactorily everything that needs to be known about this tool.

Measured drawing by Martin Bohøj. Scale 1:1

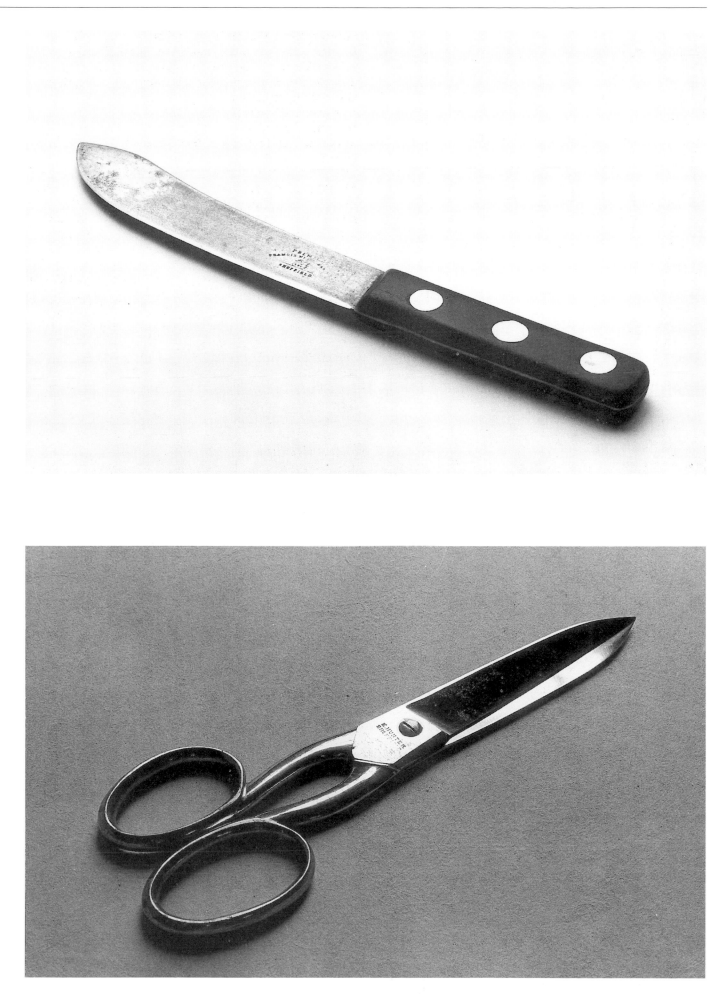

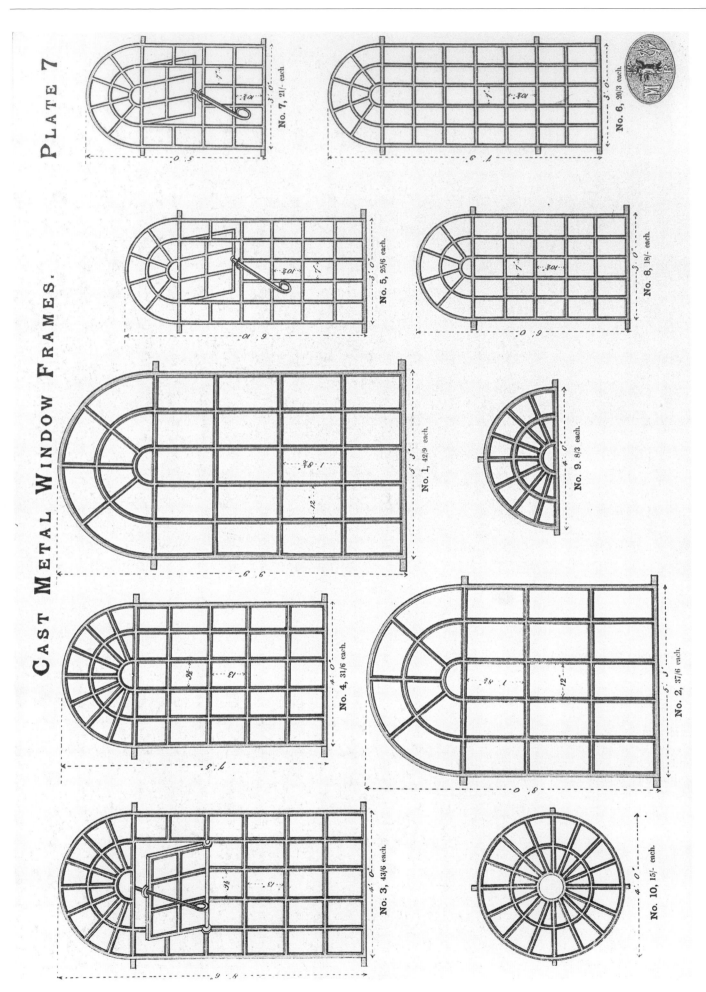

18. The Coalbrook-dale Range.

Extra Strong Kitchen Range, 32 in. high on stove or sham side, with wrought plate-iron oven, bright mounted and completely fitted; massive wrought bars, with fall-bar and winding cheek, bright falling crow, draw-out stand, bright mountings throughout; hot hearth or sham iron, or boiler front, fitted complete with coves and arch-plate, best fitted and finished, con-plete as drawing, exclusive of chimneypiece (No. 33), and *without any boiler.*

EXTRAS. { Chimneypiece No. 33, as shown ... No. 34, as page 419 No. 29, ... 417 } Cast back boiler in addition Cast L boiler instead of hot hearth

For wrought boilers, see page 411.

	4 ft. 4 in.	4 ft. 6 in.	4 ft. 8 in.	4 ft. 10 in.	5 ft.	5 ft. 3 in.	5 ft. 6 in.	6 ft.	

Price, fitted as above, exclusive of chimneypiece and boiler ... ...

	4 ft. 4 in.	4 ft. 6 in.	4 ft. 8 in.	4 ft. 10 in.	5 ft.	5 ft. 3 in.	5 ft. 6 in.	6 ft.	each.

About cast iron generally

Cast iron building details and street furniture of the 19th century still characterise city centres of the older industrial countries. Here was a technique and a material which combined to offer an extraordinarily wide range of design and constructional possibilities and which had physical and functional qualities for use both internally and externally. The casting of iron has been of prime importance in almost all aspects of industrial development and occurs in machines, products and buildings alike. Cast iron allows freedom of shape in design: the process exploits the molten metal's fluidity and ability to flow into the intricate cavities of the mould and form a complex, homogeneous component or product. Its importance lies also in its ability to withstand exposure to the weather, since it rusts much slower than mild steel, due to its higher carbon content. Historically, the importance of cast metals, and not least cast iron, is the clue the process gave to the very idea of quantity production: cast components can be repeated from the same pattern and this repetition of identical parts is the basis of interchangeability in production.

Comments on the illustrations

Many industrial countries had foundries which made windows something like these. They were a very good solution to the problem of admitting good working light to factories and workshops. The slim, cast frames and glazing bars admitted the maximum of light with minimum pane sizes, large panes being at first unobtainable and later very expensive. The catalogues of M.R. & Co., Walter Macfarlan & Co. and the Coalbrookdale Company from around the 1870s showed the extent to which prefabrication of building components was practised. Some of the other wares shown are 35 different rainwater gutter sections, 24 designs of rainwater hoppers, down pipes, boot scrapers, stair balusters, fireplaces, a vast ornamental double staircase for a departmental store, and a prefabricated cast iron railway station. Products such as this kitchen range were constructed of a combination of cast and wrought iron components according to the function they were required to fulfil. Wrought iron is tough and good in bending; cast iron is hard, good in compression, fire resistant, but relatively brittle.

Measured drawing by Martin Bohøj.

Subject: Two beam scales used on, and often fixed to, the counters of chemists, grocers etc. for accurate weighing of finer commodities.

Manufacturer: Unknown to the authors. This type of beam scale was widely used on the continent of Europe and made by a number of firms in Germany and Scandinavia.

Design: Anonymous.

Materials: Mainly brass. The weights, which are missing, would also have been of brass.

Manufacturing processes: Mainly castings turned and otherwise machined, filed, ground, buffed and polished. Lacquered with a shellac-alcohol mixture.

Dimensions: The photographed scale: height 49 cm, length 55 cm, tray diameter 18.5 cm.

Location: The drawn example is privately owned. The photographed example is at Denmark's Museum of Technology, Elsinore.

References & notes: The 1920s and 1930s catalogues of Henrik Jensen, Copenhagen, and H.O. Rasmussen, scale manufacturers and dealers, Copenhagen, Denmark. The catalogue of Georg Westphal, Celle, Germany. Avery Historical Museum, Smethwick, England.

Related products: Standard Measures pp 86-87, Ebullioscope, see p. 104.

Evaluation: Some of history's finest industrial design is to be found in the field of measuring apparatus. It started with the clock, the greatest single stimulator of mechanisation, and it continued in science and in the laboratory. Industrialisation necessitated accurate weighing and measuring and brought the equipment for doing it right into daily life. It was the prerequisite for implementation of standards, and the laws relating to these. So it is not surprising that scales for the weighing of goods by the retailer for the customer should be accurate and reliable and express these qualities in their materials, finish and design.The beam scale allows the function and the operation to remain completely obvious and open,

weights equivalent to the commodity ordered being placed on the one pan and the commodity being applied on to the other pan until the indicator is at centre. The long indicator gives a high degree of accuracy. Brass is a beautiful alloy; it is dazzling when polished, warm when steel brushed, and gets a pleasant patina when in daily use. Observation of the small, but important, differences in the construction, proportioning and forming of these two scales is instructive to the designer. The main functional difference is in the pans and their relation to the hangers: the loose pan solution on the right was to enable emptying the contents – for example spices – into a bag. It also facilitated cleaning.

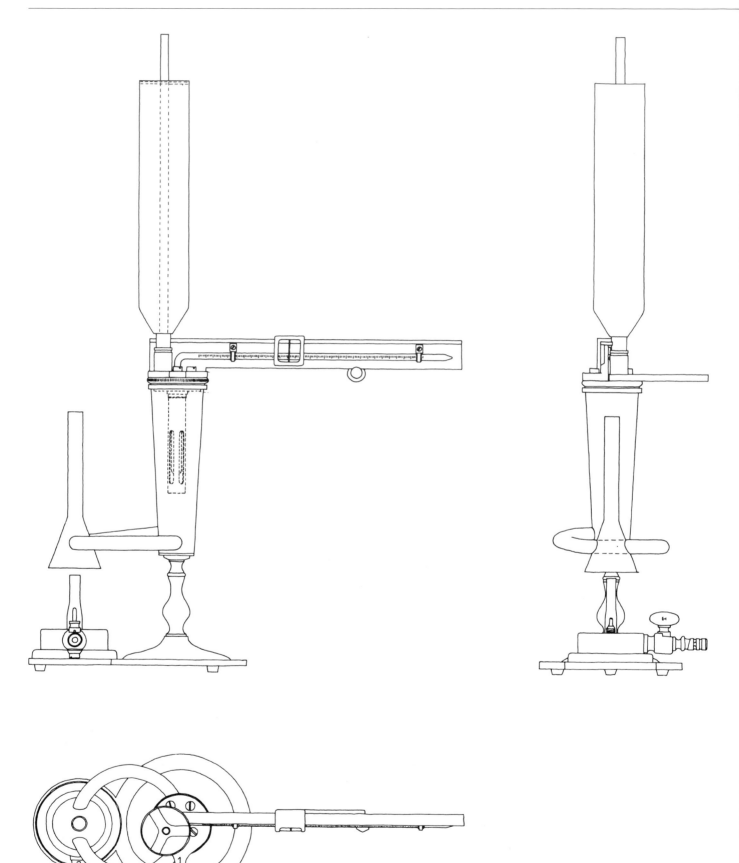

Measured drawing by Mette Milling

CM 0 1 2 3 4 5 6 7 8 9 10

Subject: ebullioscope, for measuring the alcohol content of wines, spirits and beer.

Manufacturer: This example is stamped with the name 'Struer' which is a Danish supplier of laboratory equipment. Made in France or Germany.

Materials: Mainly brass, with cast iron base painted black. Glass thermometer.

Manufacturing processes: Mostly turned and threaded castings. Polished.

Dimensions: Height 55.5 cm.

Location: Randers Museum, Denmark.

References & notes: *Brewing, science and practice* by H. Lloyd Hind, London 1948.
The 1898 catalogue of Struers A/S, Copenhagen.

Related Products: Standard Measures pp 86-87, Beam Scale pp 98-99, Thermometer pp 242-243.

Evaluation: With this instrument the alcoholic content of a liquid can be established by pouring a sample of it into the boiler and lighting the gas burner. The temperature at which it boils is read on the thermometer and it is this temperature which indicates the alcoholic content. The apparatus is designed like a tree of clearly defined parts each with its function in the measuring process: the heat source, the boiler with its heat-transfer tube and chimney, the thermometer, and the condenser at the top, all standing on a base unit. These five elements can be screwed and lifted from each other while working with the instrument or for dismantling prior to laying them in an etui. The ebullioscope was used in breweries and distilleries and by inspecting authorities. It is an exceptionally good example of a measuring instrument, which had to be accurate, simple and quick to use, robust and transportable. It is well thought out, beautifully made, and solid – qualities common to many artefacts of this period, but especially where authority and reliability were at stake.

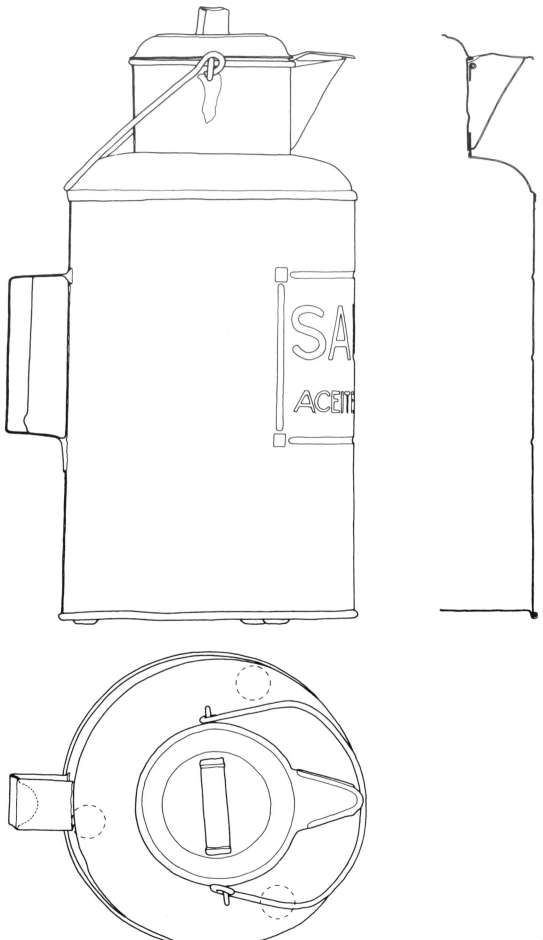

Measured drawing by Claus Bech-Danielsen. Scale 1:2

Subject: Cans for the purchase of olive oil from bulk and for its storage at home.

Materials: 0.25 mm tinplate (steel sheet coated on both sides with tin) and wire. The can left in the photo has an added galvanised steel standing ring at the base and a wooden handle.

Manufacturing processes: Tinsmith work: beading, embossing, rolling and pressing with form tools, flanging, double lap seaming, soldering.

Dimensions: Diameter 15.7 cm, height 32.7 cm.

References & notes: *Teknisk Leksikon for Haandværk*, edited by Norup, Copenhagen 1943.

Related products: Watering Can, pp 108-109.

Evaluation: We know little about the origin of these cans other than what they themselves tell us. They serve to remind us of the wisdom of the re-use container, that once important object of daily life. They are noble representatives of a whole branch of technology which is usually ignored in design literature: the skills of the tinsmith, which developed on the one hand into the food canning industry and on the other hand to the production of light, cheap utensils for a thousand different uses. The technology of the can is interesting because as a designer one 'discovers' the industry as one which has led a very

secluded existence but nevertheless embodies important skills and design ideas. The can is the most humble of all utility articles and yet is one which must have the greatest use-value for money and materials. Two things had to happen to give this trade the necessary stimulus to make the 'tin can' part of daily life the world over: the necessity to preserve food over long periods of time in varying climatic conditions – specially for long sea voyages – and the invention of tinplate. Tinplate resisted corrosion, was non-toxic, water- and air-tight and had the right strength at very low cost. Its malleability enabled the all-important lap seam joint to be made between bottom and side. The tin coating provided an excellent surface for soldering the side and top joints. Up to the 1860s all this was being done by hand. A small hand-operated rolling machine was, however, invented in about 1825 at the same time as tinplate. In the 1890s the whole process became semi-automatic. The economy and simplicity of the products shown here are their charm: the material, of which there is only one type and thickness, is very thin. Until it is jointed, beaded and seamed, it is flimsy. These few processes stiffen and put the strength just where it is most needed.

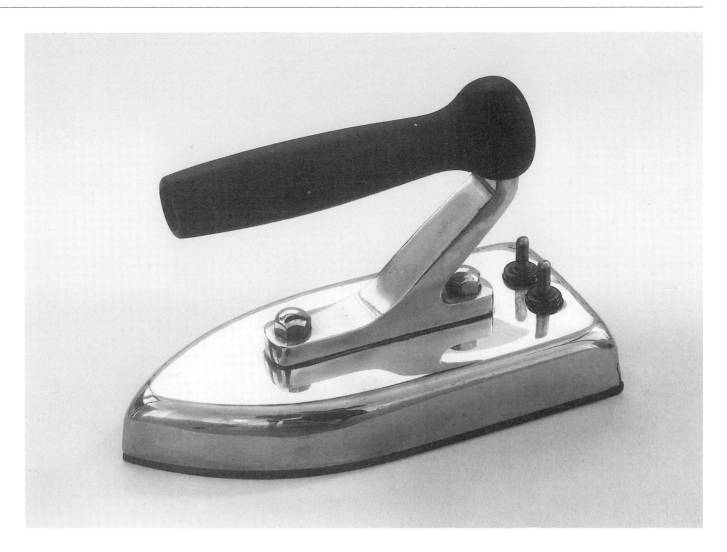

Photo of flat iron, S.R. Gnamm, Die Neue Sammlung, Munich.

Manufacturer: AEG (Allgemeine Elektricitäts-Gesellschaft), Germany. About 1910.
Designer: Peter Behrens.
Materials: Kettle: brass, with handle of rattan, lid knob of wood. Flat iron: nickel plated brass, black painted wooden handle.
Dimensions: Kettle: height 24 cm. Diameter 16 cm.
Location: Die Neue Sammlung, Munich.
References & notes: *Industrial Design* by John Heskett, London 1980.
The electrical heating element in the kettle is held in position with a bayonet-type socket, which can be seen in the photograph. This part was evidently conceived in the same way as an electric light bulb to be easily removed for cleaning or replacement.
Evaluation: These products are so clear and convincing that even now one feels some of the same amazement which people must have felt seeing them for the first time 90 years ago. They are products of the new electrical age, no doubt about that, and yet there is a kindliness in form and materials which, as with any good tool, makes you want to hold and use them. The kettle is one of a series of four basic shapes which all had rattan handles but which were available in a variety of metal finishes and treatments. The eotechnic nature of some of these treatments and the handles of natural fibres were presumably meant to make these singularly neotechnic products more acceptable to the public. However, this was probably not merely a marketing device imposed by the AEG sales department. A close study of Peter Behrens' work – a must for any industrial designer – shows that he was as interested and engaged in the crafts as he was in industrial innovation and often succeeded in combining these in his broad repertoire of design solutions.

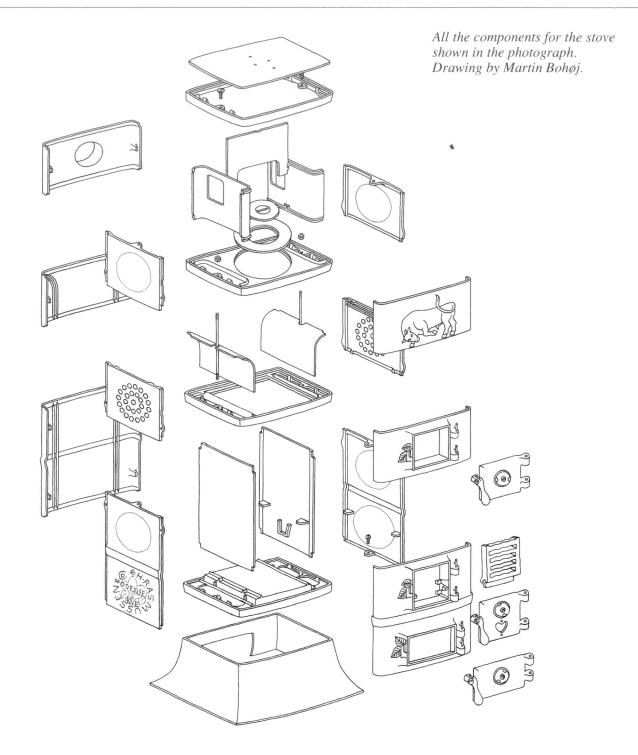

All the components for the stove shown in the photograph. Drawing by Martin Bohøj.

Subject: Domestic room heating stove series.
Manufacturer: H. Rasmussen & Co. a/s, Odense, Denmark.
Designer: Knud V.Engelhardt, 1882-1931.
Materials: Cast iron.
Manufacturing processes: Sand-casting of each component; the main stages are as follows: Hand shaping of a pattern in wood simulating the component. The forming of a cavity in fine, damp sand by pressing the sand round the pattern, one half at a time, in moulding boxes. Removal of the pattern. Fixing the two boxes together so that the impressions in the sand come exactly opposite each other. Pouring the molten iron into the cavity through access holes previously formed in the moulding boxes and the sand. Cooling, solidifying, extraction from the sand mould, cleaning off and any necessary machining, grinding, finishing. Assembly of stove with graphite putty and bolts.
Dimensions: The series is based on a module of 22 cm in height. The plinth, the double ash-and-burner module and the top module are common to the four basic sizes of stove whose heights are 100, 122, 144 and 166 cm. Plan dimensions are plinth 41 x 54 cm and stove top 31 x 40.5 cm. These dimensions do not include the water bowl for humidifying the air, which is loose.
Location: Author's collection.
References & notes: *Knud V. Engelhardt* by Erik Ellegaard Frederiksen, Copenhagen 1965.
Kakkelovne by Gunnar Biilmann Petersen, an article in *Arkitekten,* pp. 89-98, Copenhagen 1923.

Kakkelovn og jernovn by Ebbe Johansen, Copenhagen 1980.

The rampant bull in the top panel was designed by the artist Joakim Skovgaard.

Related products: Cast Iron Building Components pp 96-97, Tiled Stove, pp 174-175.

Evaluation: In the 1920s a number of Danish architects co-operated with the foundry industry to produce stove designs primarily for the heating of flats in multi-storey housing. In retrospect, we can see here a design development in which all the conditions were united to give good solutions: the need for efficient and long-burning room heaters gave rise to the tall, slim 'magazine' stove in which the fuel – coke, coal, wood or peat – gradually dropped down during combustion to feed the fire. This coincided with the need for small dimensions to save floor space in the limited room sizes; which in turn gave rise to the concept of stoves which could be built up of varying numbers of modules to satisfy different heating requirements. As can be seen clearly in the drawing, these modules gave the production and sales advantage of small, standard castings which could be marketed in different combinations. Engelhardt's solution was exceptional in that its undulating surface both increased the heating area and visually exploited the technique of casting. The design incorporated convection with cold air intake at the rear of the plinth and side ducts leading to hot air outlets (perforated) higher up. The plastic plinth form is in better proportion with the higher models than with the 4-module stove shown here.

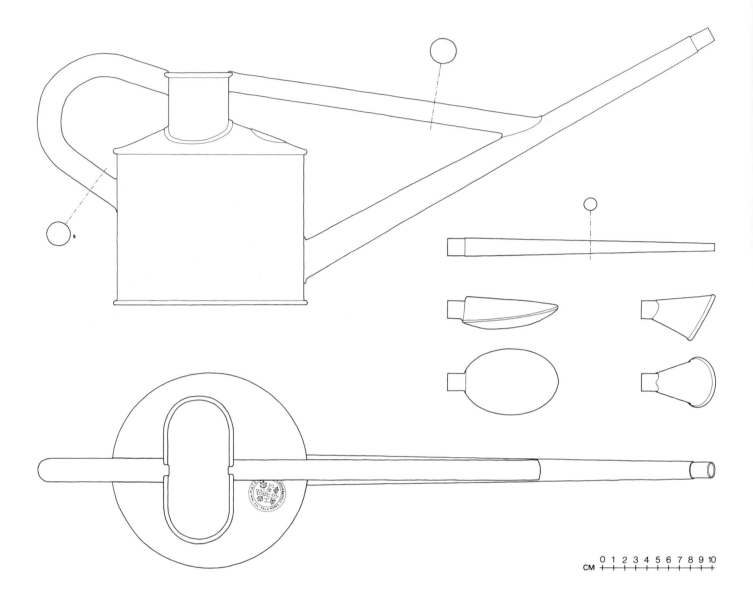

Subject: Gardener's watering can.
Manufacturer: The can shown was manufactured by Taylor Law and Co. Ltd, Stourbridge, England. The original manufacturer was John Haws Ltd. Bishops Stortford, England. At the time of writing the can is manufactured by Haws Elliot Ltd, Smethwick, England.
Designer: John Haws designed, patented and started manufacture in 1885.
Materials: 1 mm tinplate, or galvanised sheet steel, powder coat painted. Spray roses and embossed trademark are of brass.

Manufacturing processes: Tinsmith work: folding, hammering, rolling and pressing to form tools. Flanging, double lap seaming and soldering. Galvanising, painting. In the early days of manufacture the main handle was bent with the help of a lead core to prevent the tube from collapsing.
Dimensions: Diameter of water container 18 cm. Capacity 3.4 *l*. This is one of a range of sizes.
Location: Author's collection.
Related products: Re-use Cans pp 102-103.
Evaluation: By the 1930s this design was some 50 years

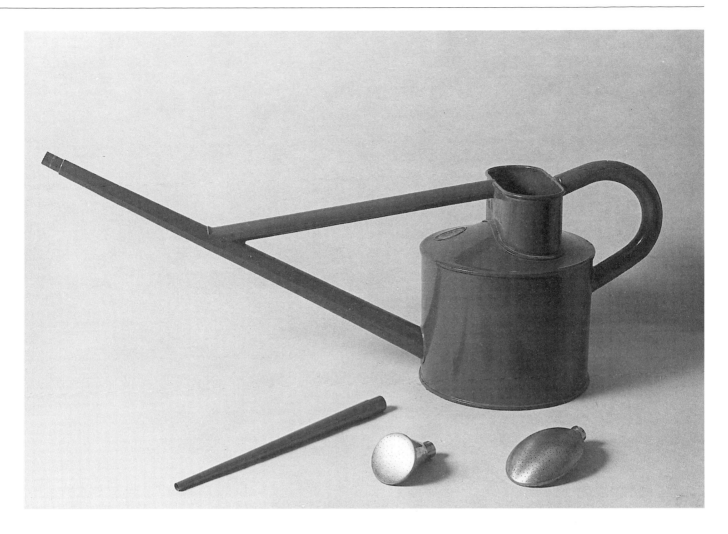

old, but we have placed it here to draw attention to the agelessness of a utensil whose design has been primarily dictated by its functions. With this watering can the gardener can reach into inaccessible plants with the help of the very long spout and extension piece; he can apply a fine gentle spray with the removable roses (an innovation in the 1880s); he can always find the right place to grip which will allow for the shift in centre of gravity while watering or filling; due to the parapet around the can opening he can carry and pour without spilling. The design is entirely based on the skills and techniques of the tinsmithing factory and the construction has changed very little since the early days of production. The can shown here was made in 1974.

Measured drawing by Karin Taidal Christensen.

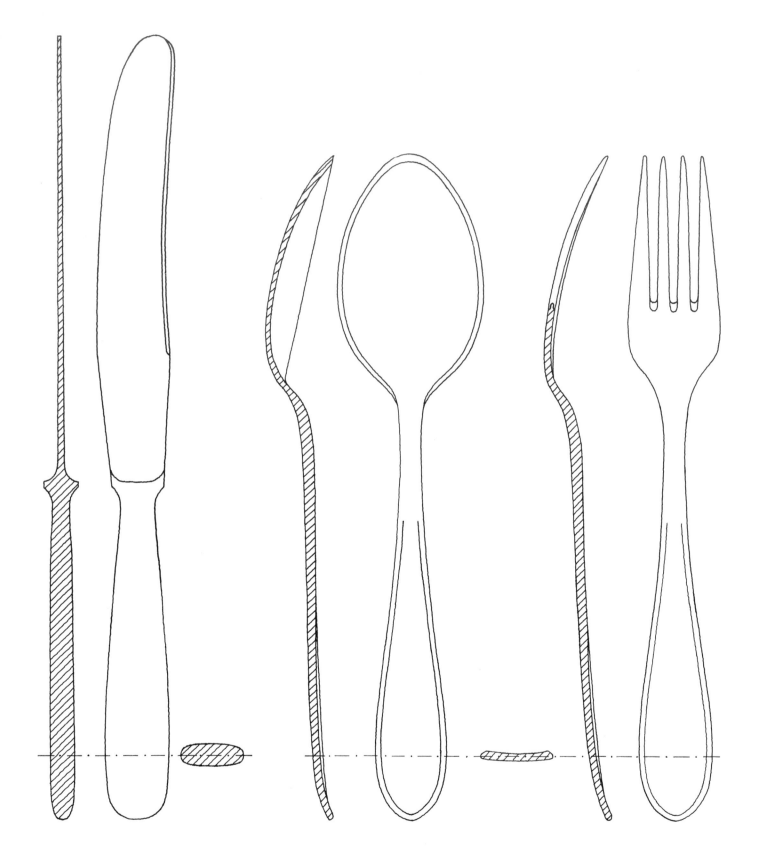

The dessert set as adopted by the Danish Standards Institute. Measured drawing by Martin Bohoj. Scale 1:1

Subject: Standard cutlery for public institutions and for domestic use.

Manufacturer: Carl M. Cohr, Sølvvarefabrikker A/S, Fredericia, Denmark.

Chronology: First produced in two knife sizes, two fork sizes and three spoon sizes for hospitals in 1929. Adopted by the Danish Standards Institute in 1931(knives) and 1950 (spoons and forks). The Cohr company augmented the series for general use and marketed it between the 1930s and late1960s.

Designer: A works design in which H.P. Jacobsen, Hans Bunde and Ejnar Cohr were involved.

Materials: Stainless steel. Alloy specification: max. 0.15% carbon, 0.25 – 0.60% silicon, 0.30 – 0.50% manganese, 17.5 -19% chromium, 7.5 – 9.5% nickel, max. 0.10% phosphorus and sulphur, remainder iron. Known in the trade as '18 – 8'.

Manufacturing processes: Machine-assisted hand die making. Cutting of 2 – 3.5 mm (according to type of cutlery) stainless steel sheet to strips. Blanking (stamping to rough form). Rolling, annealing. Initial die stamping, annealing. Die stamping of handle. Stamping of bowl or prongs. Coarse grinding, fine grinding, wire brushing, buffing.

Dimensions: See full-size drawing.

Location: Author's collection.

Related products: Cutlery for Travellers pp 76-77, Sheffield Cutlery pp 94-95, Domestic Cutlery pp 114-115.

References & notes: *Dansk Kunsthåndværk* (periodical) 1954 nos. 7-8, page 108 & 109 article by Mogens Koch. *Danish Standards Institute* specifications DS 69 May 1931 knives, DS 67 May 1950 spoons, DS 68 May 1950 forks.
Made in Denmark by Arne Karlsen and Anker Tiedemann, Copenhagen 1960.
The knives in the photograph do not belong to the standardised set. Their relationship to the set is uncertain.

Evaluation: With the exception of the knives the entire range of this cutlery, in the photograph, has been in daily use by the author's family for 40 years and we still have nothing but praise for it. Each part is the perfect tool for the performance of all functions: table laying, eating, serving, washing-up, putting into and taking out of drawer. Each is of the right thickness and weight to be suitably firm and convincing in use. The forms are comfortable in the mouth and in the hand. Other cutlery feels heavy and over-dimensioned by comparison. There is a unity of form running through the whole series. The shapes are particularly satisfactory in outline and section. The hand and eye that drew them and modelled them were guided by a very sure sense of curve and form, and a knowledge of the potential of the cold die stamping process. As can be seen from the photograph 40 years of use has left the brush finished 8/18 stainless steel almost as good as new.

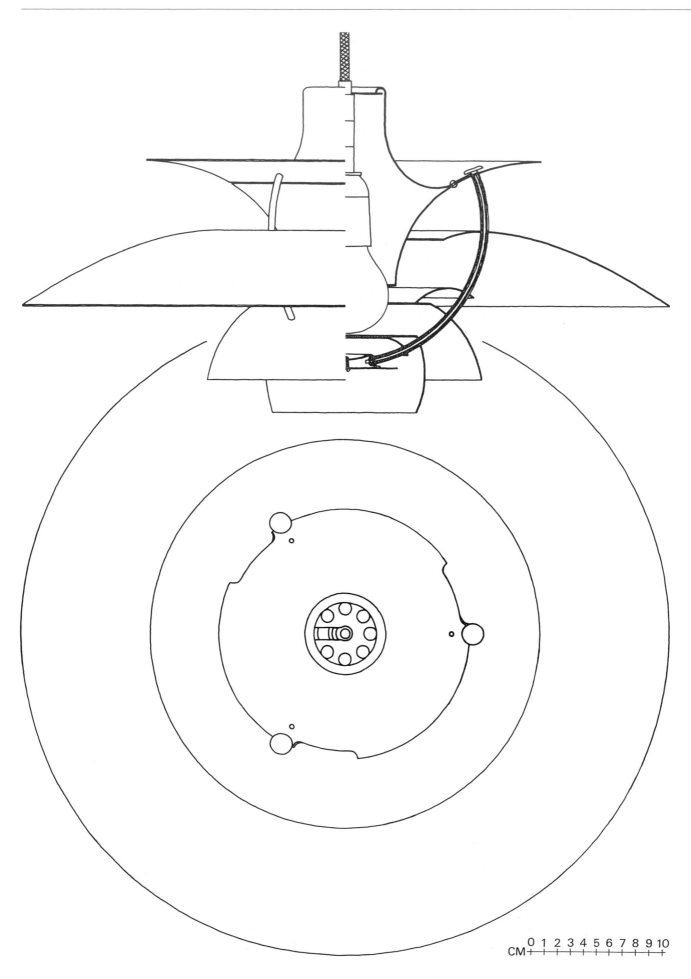

CM 0 1 2 3 4 5 6 7 8 9 10

Subject: PH5 lamp for table lighting.
Manufacturer: Louis Poulsen & Co. A/S, Copenhagen, Denmark.
Designer: Poul Henningsen.
Materials: Seven of the shades are of aluminium and the lowest one is of cadmium-plated steel. The connecting distance tubes, nuts and studs are brass.
Manufacturing processes: Stamping out the shade disks from the sheet, hydraulic pressing, boring, riveting, spinning (top shade only). Spray painted mainly white but with a system of red, blue and mauve on surfaces of some of the inner shades.
Dimensions: Greatest diameter 49.5 cm.
Location: Author's collection.
References & notes: *Om Lys*, an anthology edited by Ebbe Christensen, Sophus Frandsen, Mogens Voltelen and Steen Jørgensen, Copenhagen, 1974.
Lysmageren by Paul Hammerich, Copenhagen 1986.
Design: the problem comes first by Jens Bernsen, Danish Design Center, Copenhagen, 1983.
The recorded lamp is from 1971 production.
Evaluation: Preoccupation with the quality of light is part of the Danish culture. Light is not regarded as an abstract phenomenon which is taken for granted – something there should simply be enough of. Awareness of light penetrates painting and the crafts and is a guiding factor in architecture, and from there to daily life – night and day.

About this lamp we have chosen to quote a Dane:
'The PH5 lamp, which in its original model was designed by Poul Henningsen in 1925, employs the principle of many shades. The shapes and sizes of the shades and their placing determines the light distribution and the degree of dazzle. The colour of the light can be adjusted by the colour of the internal shades. Poul Henningsen adhered to these principles all his life but modified and refined the design to create lighting for many different functions and interiors and to correct for changes in the development of the light bulb. In this way he has produced over 40 different light fittings through 50 years of work with light. PH's lighting has come about through a chain of problem formulations and solutions and these have been a decisive contribution to the theory of artificial lighting: to the understanding of the way a room is lit, the ability of shade and shadow to describe form, the experiencing of colour and texture and what this means, and that lighting does not dazzle.' Jens Bernsen, Director of the Danish Design Centre.

Measured drawing by Ulrik Sterner Nordam.

Measured drawing by Charlotte Lemme. Scale 1:1

Subject: Table and dessert cutlery. The set called 'Provencal' consists of table knife, fork and spoon and dessert knife, fork and spoon and a steak knife. This article is concerned with the table cutlery only.
Manufacturer: David Mellor, Cutlers, Hathersage, Sheffield, England.
Designer: David Mellor.
Materials: Stainless steel blades, East Indian rosewood handles, brass rivets. The handles were later made of black, injection moulded acetyl resin which takes a pleasant wet ground surface and is dishwasher proof.
Manufacturing processes: Knives: blanked, hardened, tempered, taper ground, edged, buffed, polished, sharpened. Forks and spoons: blanked, edged, polished, formed, final polish. Handles: thicknessed, crosscut, slotted and bored to receive blades and rivets, spindle moulded to form and profile, sanded. Assembled, ground, polished.
Dimensions: See full size measured drawing.
Location: Author's collection.
References & notes: The Company's own sales material and packaging are unusually informative, clear and thoughtfully designed.
Related products: travellers' cutlery pp 76-77, Sheffield cutlery pp 94-95, hospital cutlery pp 110-111.

Evaluation: Cutlery has its own very special design problems. Each set is a family of tools in which the members have very different functions. Unlike other tools they do not only have to work well but have to look well singly, in pairs and all together. Unlike other tools they sometimes have to symbolise the place they are in, or in some way be representative. Many firms that make cutlery seem to think that their designs should try to look like more, and something other than they really are, instead of concerning themselves primarily with what functions well in the hand, with the food, in the mouth, and on the table. In 'Provencal' all these qualities have been resolved: the function and the form are inseparable. The decision to employ a second material for all handles – instead of for the knife alone, as is common – and then to give them the same size and shape, has not only given the series a strong character of its own but has unified the different parts into a set. It also reduces production costs. The curved sides and arris of the handle section make them good to hold, as does their bulk, however the latter can raise storage problems if many items are involved: in design one is often unable to achieve one advantage without losing another.

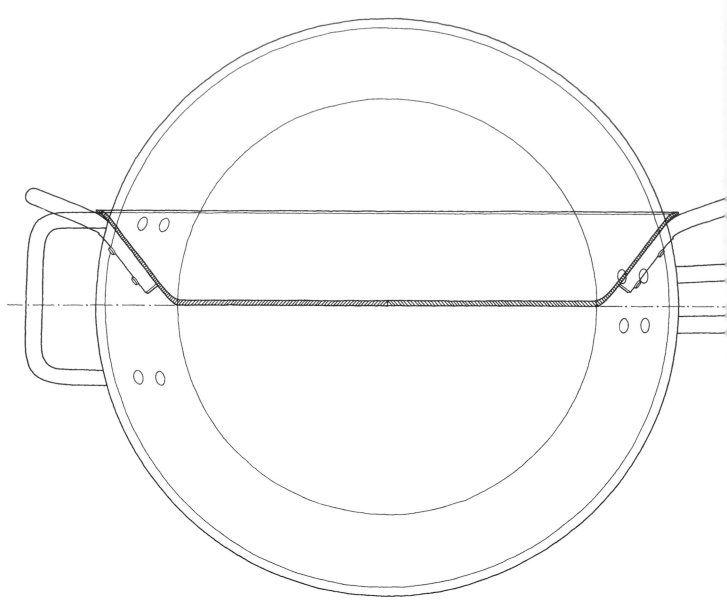

Measured drawing by Charlotte Lemme. Scale: 1:2

Subject: Eva Trio Gourmet series, here represented by a copper frying pan and an aluminium saucepan.
Manufacturer: Erik Mangor A/S, Copenhagen.
Designer: Ole Palsby.
Materials: Copper, aluminium and stainless steel according to cooking function. All handles and interchangeable lids are of stainless steel. Copper utensils are lined with a special tin alloy.
Manufacturing processes: Copper and aluminium pots are formed by spinning. Stainless steel pots are formed on drawing presses. After finishing and polishing handles are riveted to pots and welded to lids. Final finishing.
Location: Author's collection.
References & notes: Consultation with the manufacturer. At the time of writing the series has been further improved by the addition of heatproof glass lids and 4 mm aluminium bottoms to the stainless steel pots to improve heat distribution.

Evaluation: This is a series of 16 sizes of casseroles, saucepans and frying pans made in different materials in order to utilise their different heat transmission characteristics; making a total of 49 variants. The handles are something of a design accomplishment in that they perform four important functions with the utmost economy: the 8 mm diameter rod is of a thickness which is satisfactory for the hand to go round and over to the other side 37 mm away. The staple form gives a broad but comfortable handle which allows firm control with all the necessary movements while cooking. The round rod riveted directly to the pot has the minimum of contact with it so that the minimum of heat is led up the handle, added to which stainless steel is a bad heat conductor. The fourth function, that of hanging the utensil up, is beautifully obvious. The whole concept gives a very firm handle-to-pot fixing which makes the series suitable for both domestic and professional use. The pouring lip round all pots is both functional in use and serves to stiffen the rims.

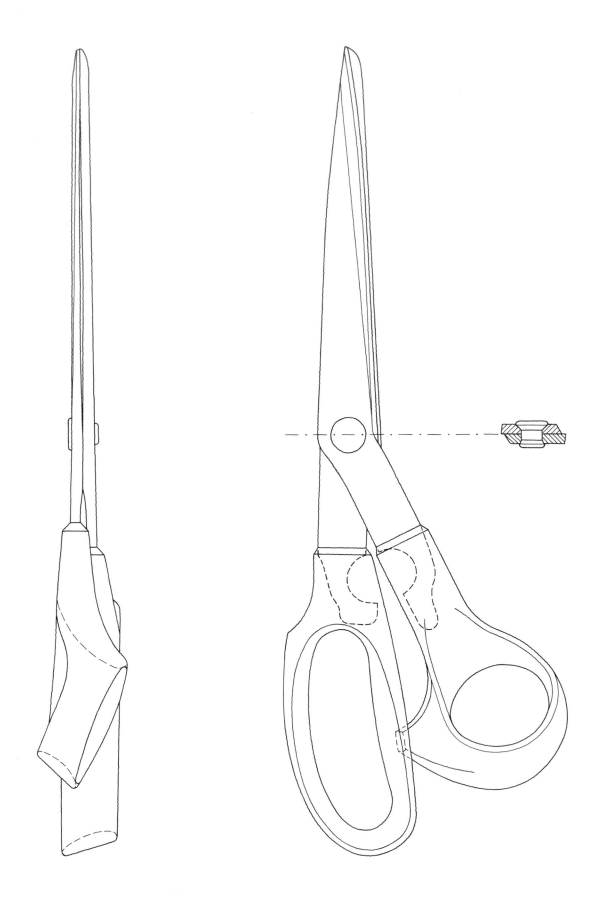

Drawing made from the factory working drawing. Scale 1:1

Subject: General purpose scissors in the 'Classic' series which comprises nine different types.
Manufacturer: Oy Fiskars A/B, Finland.
Designer: Oluf Bäckström. Product development by the engineers Rolf Lund and Olavi Lindén.
Materials: Martensitic stainless steel (chromium 13%, carbon 0.6%). Handles: ABS-plastic from 1968 to 1980. PBTP-plastic from 1980 onwards.
Manufacturing processes: The blades are blanked, hardened by heating and quenching, and ground. Handles are die cast onto the blades. Blades are riveted together under careful control to ensure the correct friction between the blades.
Location: Author's collection.
References & notes: Historical notes and early catalogue material supplied by the manufacturer. Fiskars has made scissors since 1830. The Classic series was introduced in 1968. In 1987 a new series was introduced called 'Avanti'.
Related products: Sheffield Cutlery pp 94-95,. Japanese Scissors pp 82-83.
Evaluation: Traditional scissors made entirely of steel have always suffered from a functional contradiction: either the handles are too thin and small, which can make them painful to the thumb and first finger, or the handles are generous and well-formed, as in tailors' scissors, which makes them too heavy. Fiskars used plastic, with its lightness and mouldability, to solve this problem and in so doing were one of the first to cast plastic on to a steel tool with the express purpose of improving its function. This is a tool which is immediately appealing and comprehensible. Obviously the precision ground blades will work with precision and the handles will allow the fingers to transfer the necessary power and control to the blades. Their performance is in no way disappointing. The orange handles are both delightful and easy to find in the confusion of the desktop.

The Avanti series is also an important contribution to scissor design and should therefore be mentioned here. It employs plastic ergonomically in a similar way to the Classic, but the PBTP has 15% glass fibre added which gives increased stiffness and hardness and a self-lubricating property. The material is taken right up to the pivot where it shares the load with the steel and forms the actual bearing. This is screw adjustable and demountable, an improvement on the rivet which is fixed for life. The construction is applied in various ways to the different scissors of the series, giving them greater strength, though somewhat at the expense of certain secondary functions such as slimness and visual clarity.

PART III

WOOD PRODUCTS

Wood is the only material in the book which is not synthetic, and one could question the absence of other natural materials. Why no bone, leather, or basketwork? The reason is that, much as we would have liked to include these, wood is the one natural material left which is truely industrial. It has taken a leading role, not only in teaching mankind about strength and structure generally, but also in the development of hand, and then machine tools, with which to win, work and finish the material. It was doing this long before our story starts, and will continue to do so long after it has ended. The majority of products we have chosen, with which to discuss the properties of wood in design, are chairs. By doing this we have achieved in Part III an ambition unfulfilled in the other parts, which is to concentrate on key products. But more important is that chairs can tell us most of what we need to know about construction in wood. Within the functional tradition, four chair types stand out: the Windsor chair of England and USA, the Shaker chair of USA, the Thonet chair of South East Europe (beginning in Austria), and some recent Danish chairs. These four types are instructive to compare because they each have a constellation of factors which differ greatly from one another, and which clearly show in their very different constructions: the Windsor chair, an invention of the woodsman and subsequently a product of industry; the Shaker chair, a product of a religious sect, who had their own and original functionalism as part of their religion; the Thonet chair, the creation of a brilliant father and his ingenious sons, who made an industry based on a quite new chair construction and manufacturing technique; and Danish chairs, particularly of the mid-20th century, which do not have a common construction, but have been influenced by the previous three chairs, as well as by others. They are the result of a few designers and makers applying their skill, experience, and respect for wood to a national interest, the equipment and design of buildings. Each of these chair types have their own special physical and aesthetic qualities and there is a clear honesty about them all. They are all functional within their own terms. They have all, in their day, been used by large numbers of people because they have been robust yet inexpensive.

Our history of industrial design starts in an environment consisting largely of handworked wood. Even Christopher Polhem's advanced, water-driven machinery was almost entirely wood. Metal, principally iron, was reserved for the critical working parts, and for holding the wooden parts together, and to make tools for working the wood. We can see this situation beginning to change in the products and machines of the early 19th century: the railway wheel is an example. It is made of wood and iron but the iron takes the dominant role. It is the only example in the book of that important, and long lived, stage in the history of technics when wood and iron worked in close partnership in the ingenious, often beautiful, designs for highly stressed structures. Other examples are the horse drawn vehicle, and farming equipment.

When considering the material wood, it is worth thinking first of the great plant from which it comes, the tree. We have become so far removed from nature that we tend to regard wood in terms of planks or table tops – simply material, squared off, straight, and flat. Just as trees seem endless in their variety of form and detail, so is wood many different types of material. It has such an enormous range of physical properties and colours, that just to call it 'wood' is simplistic. *Wood is a whole group of materials.* In many languages the word 'wood' does not exist. It is called 'tree'. While trees are growing they are using and absorbing carbon dioxide, the principal greenhouse gas, as well as performing many other positive environmental functions. They are an indispensable component of nature: both a continual process and, when properly forested, a permanent source of material. In this way wood is unique in maintaining a balance in the living environment. In the same way that the fully grown tree itself is a formidable structure, capable of withstanding the enormous loads imposed by the elements, so is wood capable of almost any structural load required of it by the engineer, the architect, and the designer.

Wood is not only universal in the uses to which it can be put, but it can also be formed, worked and manipulated by the simplest of hand tools and human energy. It introduces us, as children, to the feeling for making things and, if we wish, we can go right through life experimenting and designing in wood. Wood lends itself so well to clean cutting by sharpened steel that it has led the way in the development of the machine. Good examples of this include the machines for making pulley blocks discussed in Part I, and the block itself here in Part III, as well as Blanchard's copying lathe (Part I). Research continues to discover new aspects and uses of wood within machining, forming and drying techniques. The latest we have heard of is an air drying method which reduces the large amount of energy that goes into the kiln drying of timber.

If wood is so admirably suited to so many uses, why has it lost so much ground to metals and plastics? This is a complex matter which is concerned with economics and inadequate afforestation policies. In addition, if we concentrate on the material itself, it has three negative properties. All materials are – more or less – unstable, meaning that they change their dimensions slightly when subjected to heat and/or moisture changes. This is important for the designer to remember. In most woods these changes are quite pronounced. It swells in increased dampness and contracts in increased dryness, and this movement is accompanied by changes of form such as warping and twisting. To reduce movement and contortion as much as possible, the converted wood (sawn to planks) is dried so that its moisture content is in equilibrium – or nearly so – with the moisture of the air in which it is to be used (exceptions to this practice are mentioned where they occur). In the course of history, the interiors of buildings have become better heated, so great care with the wood drying (or seasoning) process is of the utmost importance. It can be said that wood is a *living* material; it never stays completely still, it moves with the seasons of the year, indoors and out.

The second point to remember about wood is that unlike the other materials discussed, it has a dominating grain; it has a directional structure up and down its stem, or trunk, as do nearly all plants. This is fortunately very clearly seen in most woods, and is one of its greatest attractions, but it does mean that a board, for example, is strong in bending along the grain but weak in bending accross the grain. It splits along the grain but not across the grain. The visual, and physical, difference between side grain and end grain when under the tool is so marked that this two-fold characteristic quickly becomes accepted as part and parcel of the material.

Thirdly wood is burnable, and this has obvious disadvantages. Though sitting writing on a cold November day, one is glad that it can burn in the stove and keep the house warm.

For much of the story related in this book, all this has been well known. Wood has been the best known material of any, and indeed, understood by a much greater proportion of the population than today. But industrial research in the 20th century has done much to physically stabilise wood. Sheet materials such as plywood, blockboard, chipboard (particle board), and hardboard are some of the products which reduce movement in wood. In short, they work on the principle of taking the 'life' out of wood by cutting it up, and then gluing it together again under pressure to form large panels of various thicknesses. These methods have changed design practice in joinery and case furniture considerably. For example, in the 1930s the fairly stable flush door became a possibility, and it was no longer necessary to surround solid wood panels with a restraining frame.

Finally, a word about the measured drawings in Part III. In all matters to do with design, seeing is believing. The drawings are faithful and accurate recordings of individual artefacts made in full size, directly from the subjects, but often reduced for the book. They embody a great deal of design enlightenment because they show real thicknesses, real lengths, actual curves, and if read with the photographs, show what all sorts of sections and profiles look like 'in the round'. In the measured drawings can be found basic dimensions such as seat heights, depths and widths, and the sort of chair leg dimension necessary to bear the weight and movement of a heavy person. To the unaccustomed, some of the drawings are difficult to understand, and need a little time and practice to get used to. They look complicated because the three projections – plan, side and front elevation – are drawn overlapping. This is so that, when drawing, one can get all three projections in full size, on to one large sheet of paper, and this makes for much greater accuracy, as well as helping the drawing process. The chair to be measured is carefully placed and chocked up on the drawing paper, first standing on its legs, then lying on its side, and then lying on its back. In each position the chair parts are transferred down to the paper with the help of plumbline, T-square, set-square and direct marking. Pencil dots are made on the paper to mark these points. Where there are curves in the chair many closely spaced dots are necessary. These are then joined by lines which show the exact forms and structure of the chair. It will be noticed that the lines are drawn freehand; this is because it is important that it is the subject's own forms that are recorded, not those of linear drawing instruments. Hidden joints and other details which the draughtsman cannot see are not shown. While doing all this one is learning more about the chair, or object, than a hundred photographs could tell. In the interest of clarity some of the drawings show the three projections separated. Known scales have been used where possible, otherwise measurements can be taken off the drawing with the help of the scale lines.

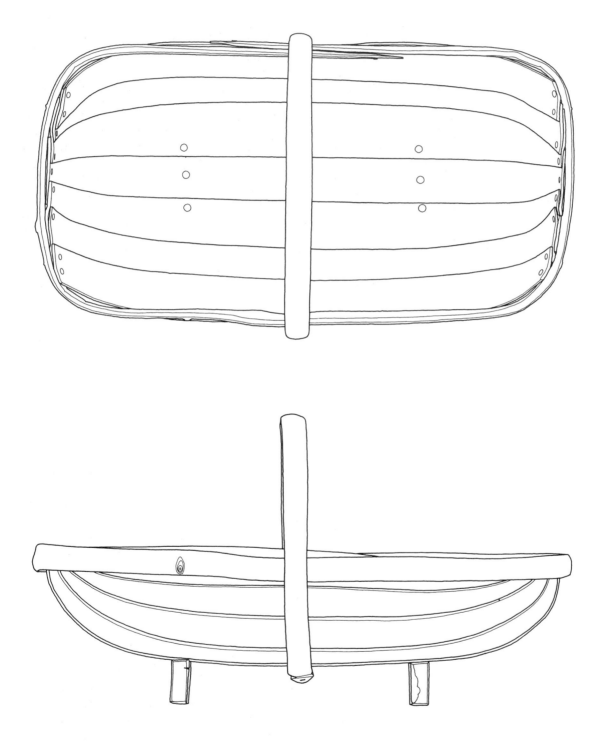

Measured drawing by Charlotte Lemme. Scale 1:4

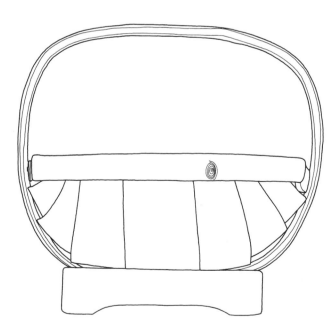

Subject: A basket for carrying garden produce.
Manufacture: This example is made by Smiths, Hurstmonceaux, England. Early 1970s.
Design: Vernacular.
Materials: Traditionally hazel frame and handle, chestnut or oak slats – all of which split well. Copper nails (which do not rust) for clinching. This example has slats of poplar which is light and tough but does not split well and is therefore sawn.
Manufacturing processes: Splitting and bending of wood while still 'green', sometimes with the help of steam or hot water, fairing, springing slats into place and nailing and clinching. Trimming with knife. Mostly hand work done on the lap.
Dimensions: Width 31cm, length 55 cm.
Location: Author's collection.
References & notes: *The Forgotten Arts*, by John Seymour, London 1984.
Evaluation: Even in industrial countries, well designed, pre-industrial, artefacts are still being made and used today. The trug has an obvious affinity with the ancient clinker boat construction; the Danish word for 'trough' is 'trug' – both indications of ancient origins.The trug is useful because it is openly accessible, accommodating, light and holds and carries well. It achieves these qualities with a construction which is quick to make, very economical and very strong due to an optimal disposition of materials: Its function, appearance and construction are as one.

Subject: Comb-back Windsor chair *c*.1760.
Maker & designer: Unknown, English.
Materials: The chair is painted in a subdued green which makes wood identification difficult but it is possible to see that some of the back spindles and the arm-bow are ash, the seat elm and the legs oak.
Making processes: Sawing, splitting, shaving, planing, moulding, turning. Bending with the help of boiling water or steam. Adzing to get seat dishing. Boring. Scraping. Assembly. Finishing.
Dimensions: In the absence of a measured drawing here is a list of measurements in cm.
Heights: Back top rail to floor, max. 107.8. Back top rail to seat (parallel with back) 74.0. Seat to arm (front left) 30.0. Seat to floor (front corner right) 40.4. Seat to floor (lowest point right) 37.5. Seat to floor (rear) 36.0.
Widths: seat front 60.7. Arms (max. external) 68.4. Arms (max. internal) 60.7. Back top rail 69.8. Front legs (ext. at seat) 48.7. Front legs (ext. at floor) 61.5. Back legs (ext. at seat) 41.5. Back legs (ext. at floor) 60.1.
Thicknesses: Seat max. 4.8. Legs thickest diameter 3.8 – 3.9. Smallest diameter 3.2.
Location: Victoria and Albert Museum, London.
References & notes: *Welsh Stick Chairs*, by John Brown, Newport 1990.
The History of Chairmaking in High Wycombe, by L.J. Mayes, London 1960.
The Windsor Chair, by Ivan Sparkes 1975. Article in *Arkitekten* (Danish) nr.11, 1982, by the authors.
Related products: Double Bow Windsor pp 128-129.

Evaluation: Considering their intricacy it is surprising that the Windsor chair is geographically so widespread. They were made in several parts of Britain (though mainly in High Wycombe near Windsor), in Scandinavia and in the USA. The Windsor is the most expressive example of the vernacular we know. Its various forms and expressions – and there are many, as we will see in some of the following pages – are mostly due to its functional and structural idea, and to the conditions set by making; not so much to style or fashion. As with so many eotechnic inventions, the Windsor chair was a brilliant piece of design. The conception of a saddle formed seat which, with its great thickness, firmly holds legs bored into its underside, and a half circle of thin spindles, or sticks, bored into its top side is a very interesting solution. The arm bow and the top rail assemble the spindles to a firm, curved structure which in spite of its slightness is so robust that it can be leaned against heavily. The making of this structure requires skill and judgement and an understanding of the properties of various woods. An example of this is the choice of elm for the seats. The seat is the structural foundation of the Windsor chair. Elm does not split easily and is therefore able to resist the splitting tendency of repeated hole boring at close centres to take legs and spindles. The multi-directional grain structure of elm also enables it to withstand the bending loads applied by the weight and movement of the sitting person. Added to this, elm feels warm to sit on.

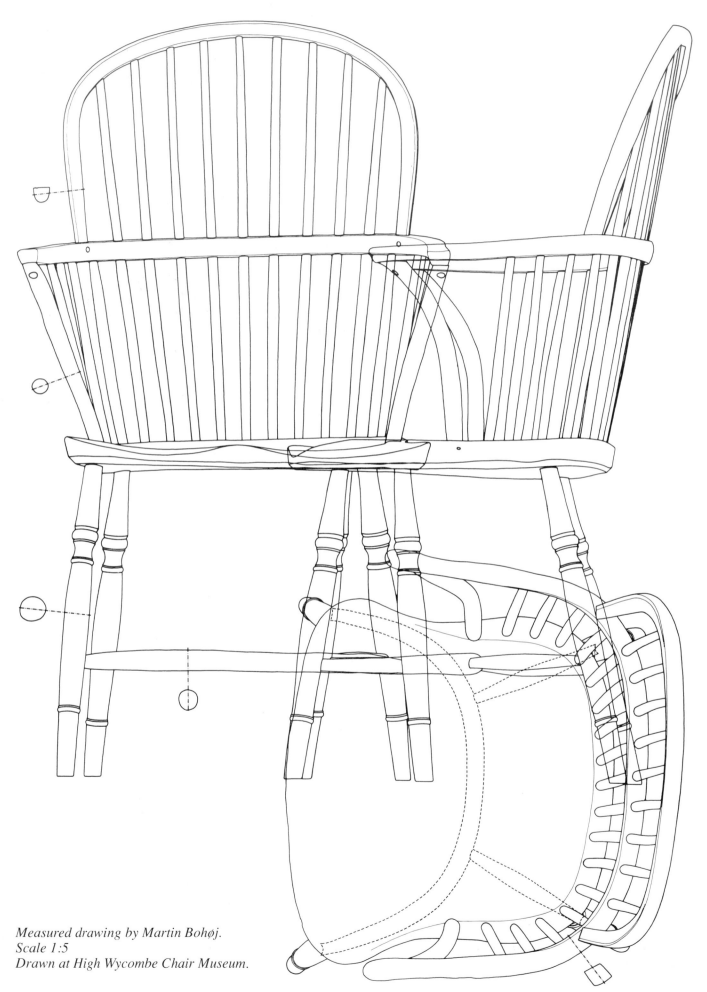

Measured drawing by Martin Bohøj.
Scale 1:5
Drawn at High Wycombe Chair Museum.

One of the two classic Windsor chair constructions. This is such a beautiful example that we have decided to let the measured drawing and the photo alone speak for themselves. The materials are almost entirely yew, except for the seat which is elm. See pp126-127, 148-149, and 150-151 for further relevant information.

Stages in production: The shell

The sheave

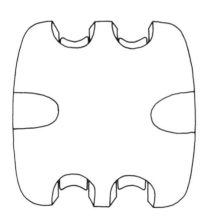

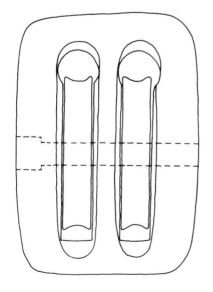

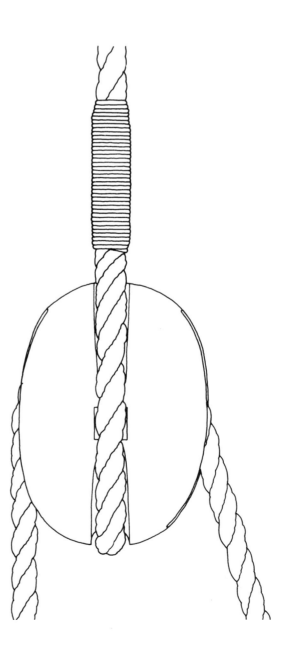

Scale 1:2

Assembly

Subject: Single and double pulley blocks.
Manufacturer: The Block Mill at the Naval Dockyard, Portsmouth.
Design: Vernacular, but Walter Taylor introduced the sheave bearing detail.
Materials: Outer shell common elm, sheave lignum vitae, axle pin forged iron, coak (or bearing) cast bronze.
Manufacturing processes: Shell: mechanised sawing, boring, mortising, shaping, moulding, polishing. Sheave: mechanised sawing, rounding and boring, turning, routing, riveting. Pin: forged and turned. Assembly.
Dimensions: The drawn example is 13.5 cm long.
Location: Author's collection.
References: *The Cyclopaedia of Arts, Sciences and Literature*, by A. Rees, London 1819.
The Portsmouth Blockmaking Machinery, by K.R. Gilbert, Science Museum, London 1965.
Early Industrial Design in England, by Adrian Heath and Aage Lund Jensen, Arkitektskolen i Århus, Denmark, 1974.
Related products: Blockmaking Machinery, pp 34-35.

Evaluation: The pulley block is a fundamental part of the running rigging of the sailing ship and at this period some thousand blocks of different sizes were used aboard the larger vessels.

The most remarkable aspect of the blocks shown here is the machines with which they were made. This is discussed in Part I. But as a piece of functional design the blocks themselves are worth studying. The shell is made of elm because this is an exceedingly tough wood which can take a lot of punishment and will not easily split under the tremendous loads to which it is exposed. The sheaves are made of the hardest and most homogeneous of all woods, lignum vitae, which when sawn in slices across the grain retains the concentricity required by its pulley function. The sheave bearing, or coak, is simply a lugged bronze tube with an internal, spiral grease channel, rotating on a forged iron pin, each with accurately machined bearing surfaces. The whole simple assembly is well-rounded, to prevent it catching on the rigging, and circumscribed and bound by a spliced and lashed strop which, in addition to holding the block, also secures the axle pin.

Measured drawing of a 4 in., 2-sheave block made with the Portsmouth machines. This block had not been completely finished. Drawn by Nis Øllgaard. Photos: British Crown copyright, Science Museum, London.

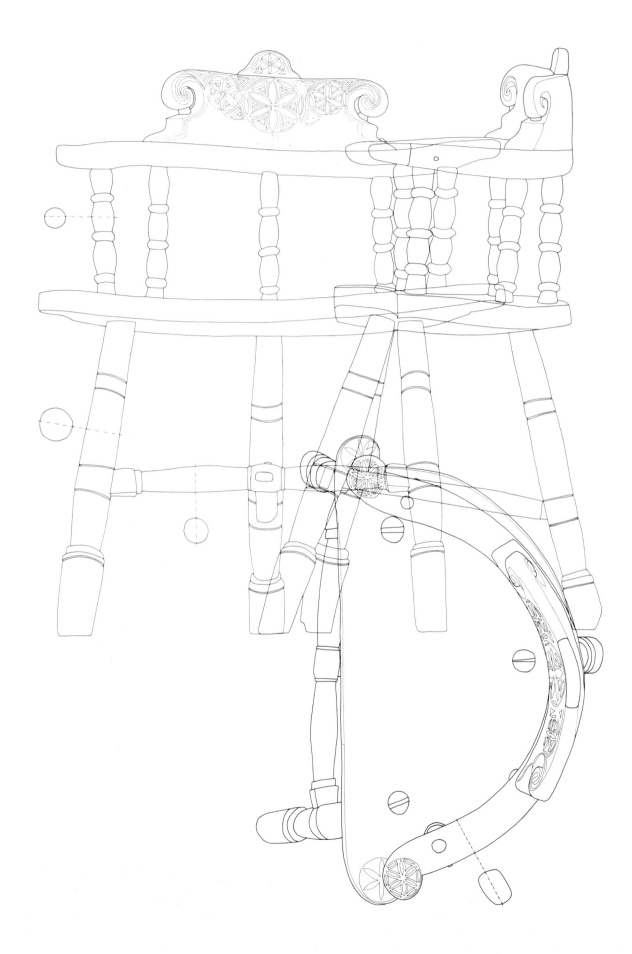

Measured drawing by Aage Lund Jensen. Scale 1:5

This chair can hardly be described as an example of industrial design but we have included it for three reasons: while working with it we began calling it the 'Danish Windsor' because of its structural relationship to the Windsor chair; that is to say, legs from below, and back supports from above, all bored into a slab seat. It describes, without words, the probable design origin of the Windsor chair: if you remove the back you are left with a sort of high milking stool. Thirdly it is a good concept which deserves to be taken up again. It is made of beech wood which has been painted deep red. The legs and balusters are carved, to imitate turning. The three legs ensure stability on uneven floors. It is from the early 18th century and is now at the Open Air Museum, Den Gamle By, Århus, Denmark.

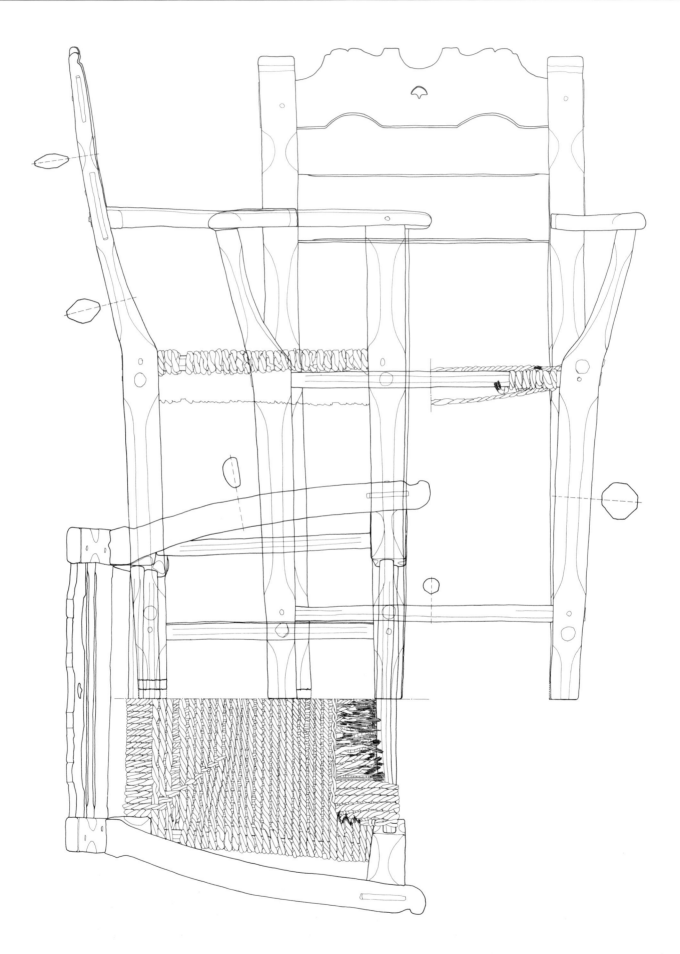

Measured drawing by Akis Nakagawa. Scale 1:5

Subject: Chair with arms for general use from the first half of the 19th century.
Maker/designer: Unknown/vernacular.
Materials: Legs, arms and back of beech. Stretchers and seat rails of ash. Painted a mahogany colour. Seat of twisted straw.
Manufacture: Handmade. Judging by the curved members and numerous jointing angles this is the work of a chairmaker. Consistent use of through joints with end grain of mortices showing, cross dowels throughout. The design and making indicate that the chair was one of the maker's standard models and made with its own set of templates.
Main dimensions: Height 86.0 cm, seat height 44.0 cm, greatest width 57.0 cm, greatest depth 47.5 cm.
Location: The Open Air Museum, Den Gamle By, (The Old Town) Århus, Denmark.

Evaluation: With very simple means this chair welcomes the user both visually and with its actual comfort. On first sight its form and construction are familiar to the Dane coming as it does from a long line of country chairs which were made on the farms. But this chair is so well-planned and consistent in both shaping and jointing that it must be the work of a good chairmaker. Adjacent stretchers and rails enter the legs at different levels to allow strong jointing. The generous chamfering of the legs pays respect to, and accentuates, the wisdom of the construction as well as visually and physically lightening the chair. As was quite common with vernacular chairs the top back rail was reserved for a touch of 'fashion' which often looked out of place and lacked the conviction of the rest of the chair.

Railway wheels on their axle. Wheel diameter 97 cm, depth of hub 21 cm. From a measured sketch.

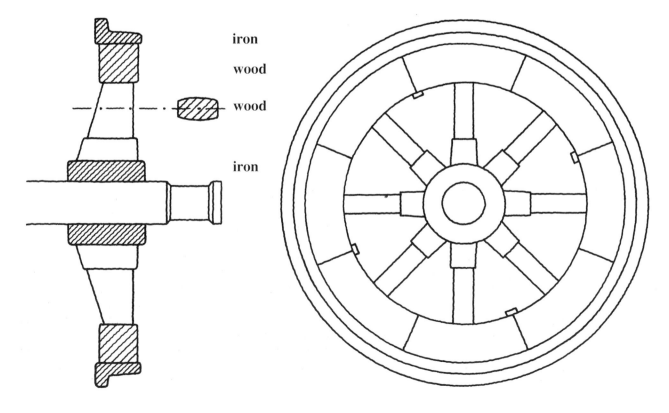

iron

wood

wood

iron

Wortley Forge dates from 1640 but the oldest existing hammer, above, is from the early 19th century. The very 'advanced' looking water wheel of cast iron can be seen to the right. The wooden paddles are missing. Production of the forge has also included edge tools and the processing of iron.

One of the experiences during this research which has left the deepest impression was our visit to Wortley Forge, near Sheffield. This was the nearest we have been to 'treading into' the early industrial environment. The whole atmosphere of the works expressed the struggle to change production methods to cope with new demands. There are two large and extremely heavy water-powered hammers which had been used, amongst other things, for forging wheel axles for the railways. The hammer shown in the photograph has a huge ashwood spring which is held over the cast iron hammer shaft. When the hammer is lifted, it presses the spring upwards. When the hammer is dropped, the spring exerts extra power to the down-ward blow. This sounds like a slow process but in fact the lifting and dropping was repeated more or less rapidly according to the rate of rotation of the water wheel and the forger's requirements.

Standing in the same large space is a pair of railway wheels on their axle. These are also made of cast iron and wood. Both the machine tools and their products demonstrate forcefully the very act of transition from a technology of wood to one of iron. Train wheels based on this rather surprising composition of wheelwright and foundry work were made in great quantities throughout the early railway age both for home and foreign railways. Compare with the steam hammer, pp 40-41.

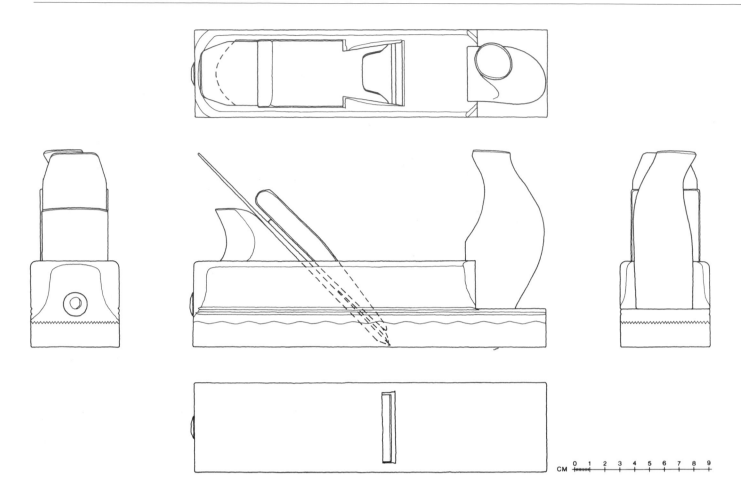

Smoothing plane by J.P.B.O., Denmark, 1950s. Measured drawing by Charlotte Lemme.

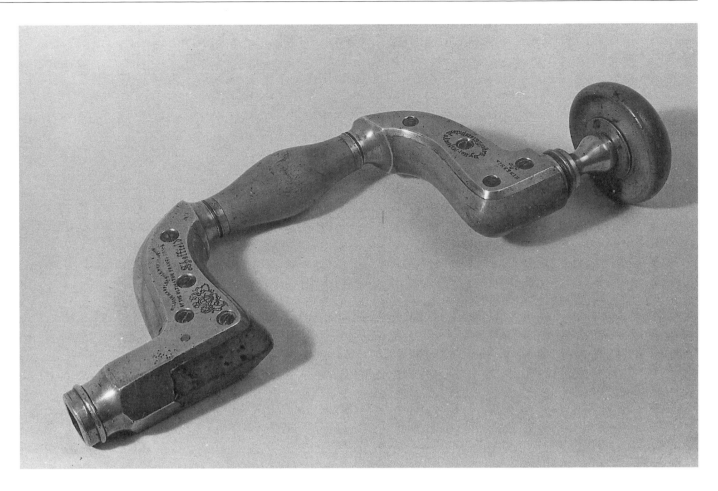

'Ultimatum' brace by Marples, England, 1850-1900.

Geared drill in steel, brass and hardwood, French, 1st half of the19th century. Photo: Science Museum.

The history of the industrialisation of hand tools starts well and ends badly. Tool manufacture as an industry was at its best when these tools were made. Their manufacturers had inherited the experience of craftsmen who made their own tools to suit the different functions of the job. Tool makers' catalogues had page after page of the same type of tool with variations appropriate to the particular work. These tools were made largely by hand in the same high quality as that prevailing amongst cabinet makers, chairmakers and joiners. With the increase of mechanisation hand tools lost their central role and so their variety and quality. Today we have become used to a complete change of situation in which a single, usually rather poor, tool has to be used for a variety of jobs.

The type of plane shown here was familiar on the continent of Europe throughout the 19th century, and is still made. This example has a stock of Utilie mahogany, a sole of hornbeam and handles and wedge of service tree. The joint between stock and sole is made with a zig-zag cutter diagonally. The brace is made of brass, steel and boxwood. The drill has handle shapes which are a study in pressure, holding and grip; in fact, all three tools exemplify what could be called the ergonomics of experience. See The Copying Lathe, pp 38-39.

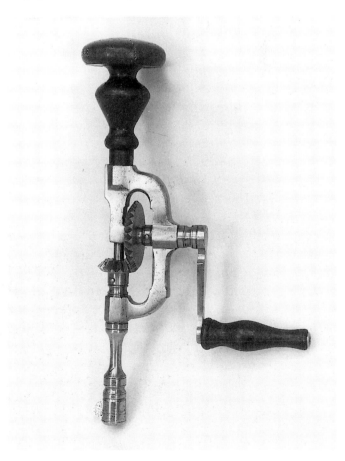

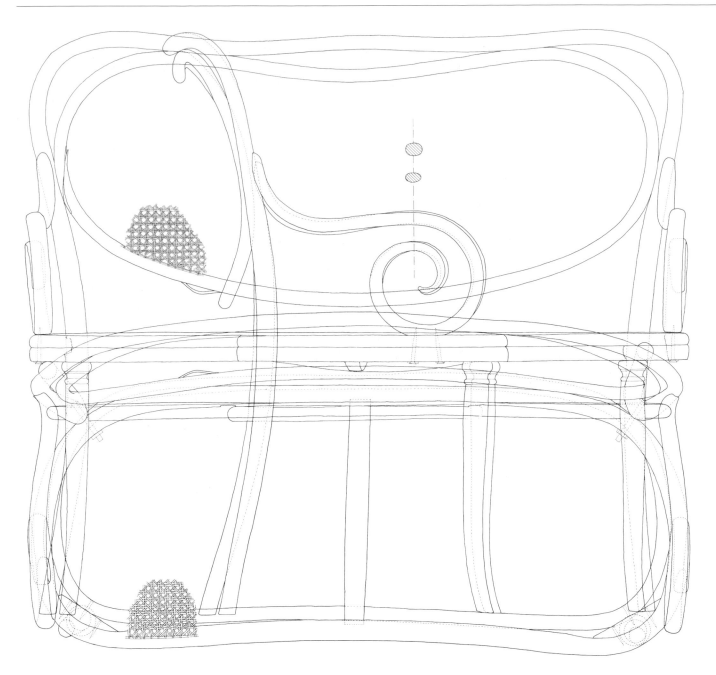

Measured drawing by Karl-Göran Malmvall.

CM 0 1 2 3 4 5 6 7 8 9 10

Subject: Settee, catalogue No 3.
Manufacturer: Gebrüder Thonet, Austria.
Designer: Michael Thonet.
Materials: Beech wood and rattan. Iron screws and bolts. Dark brown stain.
Manufacturing Processes: Converting from the log, air drying. Turning, steam bending in form-tools, cooling, drying and finishing. Spindle moulding to profile (seat frame). Boring, assembly with screws and bolts. Sanding, staining, lacquering, caning. See Chair 18 on next page.
Dimensions: Height 100 cm, seat height 48 cm, length 113 cm, depth 58 cm.
Location: Den Gamle By, (The Old Town) Århus.
References & notes: *Das Biegen des Holzes*, W.F. Exner, Leipzig 1922. *The Roots of Modern Design*, H. Schaefer, London 1970.
Historien om en stol, Ole Bang, Copenhagen 1979.

Gebogenes Holz, Asenbaum & Hummel, *Das Thonet Buch*, A.von Vegesack, München 1987.
The settee shown here is one of a pair delivered to the Elsinor Theater in the 1890s.
Related products: Thonet Chair Nr.18, pp 142-143. The Thonet Chair Company, pp 46-47.
Evaluation: The fascinating Thonet story cannot be told here. One can get a very good insight to this major piece of industrial design history from the references named. But if some of the story should be told through a single Thonet product, there could hardly be a better one than this settee. Contrary to common belief, when Michael Thonet started designing and making furniture it was with the pressed lamination technique and not with the steam bending of solid wood. This settee, therefore, has a laminated forerunner in the settee, with chairs, which were shown at the 1851 World Exhibition. Although the

Laminated settee with chairs from the catalogue of the 1851 World Exhibition.

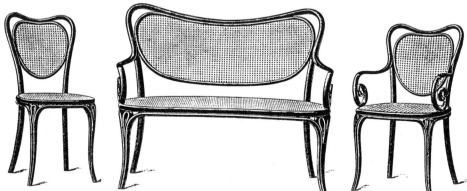

forms are similar, the structural principle is different. The earlier laminated model has a rigid, glued construction whereas model No 3 goes wholly over to the pin joint principle. A high degree of rigidity and remarkable toughness could in this way be achieved due to the long, continuous, highly flexible steam bent members which become rigid when simply connected with a bolt here and a screw there where they cross or touch each other. The ingenuity of this completely new way of making chairs, which the world needed, at a low price, and in very large numbers, was the foundation of the Thonet success.

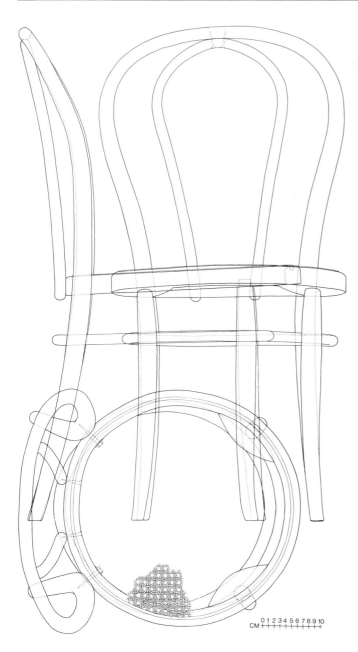

Measured drawing by Birgit Luxhøj Nielsen.

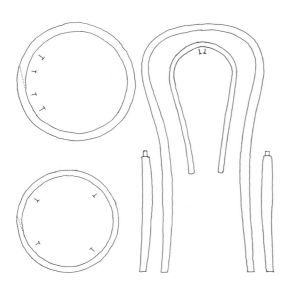

Subject: Thonet chair, cat. no.18, 1867. The drawn chair is of later date and made in Czechoslovakia.
Manufacturer: Gebrüder Thonet, Austria.
Designer: Michael Thonet.
Materials: Steam bent beech wood, seat of rattan or formed plywood.
Manufacturing Processes: Log sawn to 30 mm thick planks and then to scantlings of 30 x 30 mm and 30 x 40 mm (for seat ring). Scantlings cross sawn to required lengths. Air drying. 30 x 30 mm lengths turned on template lathe. 30 x 40 mm lengths profiled on spindle moulder. Parts placed in cylinders and pressure steamed to 125°C. Parts immediately bent to required form, with steel band outer bend-restrainer, against cast iron mould and cramped. Parts heat dried in their moulds. Parts removed from moulds and sanded. Seat rings and stretcher rings scarf jointed and drilled for leg and back joints. Parts lacquer stained. The 6 parts assembled with 10 screws or packed in boxes for assembly at destination. Parts for 36 chairs occupied 1 cubic m.
Dimensions: Height 90 cm, seat height 47 cm, seat diameter 41 cm.
Location: The chair in the photograph is at Die Neue Sammlung, Munich. The drawn chair is from a private collection.
References & notes: For references see settee no. 3 on the previous page. Thonet was the first to produce pressed, formed laminates, here as seats.
Related products: Settee no. 3, pp140-141. The Thonet Chair Company, pp 46-47.
Evaluation: Seldom in the history of industrial design have industry and design been integrated to the extent they were at the Thonet factory. The structural principle on which the designs were based, although very simple, was a formidable innovation which required the invention of a special production method. Apart from being entirely steam bent in special forming tools and having, for a chair, a new type of jointing, the designs raised completely new production problems. To take one of them, which is seldom discussed, the difficulty of handling and machining very long thin members and the uniform timber quality they demanded; the back leg bow of chair no.18 is 204 cm in length. Chair no.18 is not quite so elegant as the famous no.14 but the inner bend in the back provides slight, though important, lumbar support. It was produced in millions and exported all over the world.

Left: making is separated from assembly for the first time in chairmaking history. Parts are standardised and interchangeable both within a model and between different models, i.e. five of these parts are used in other models.

A

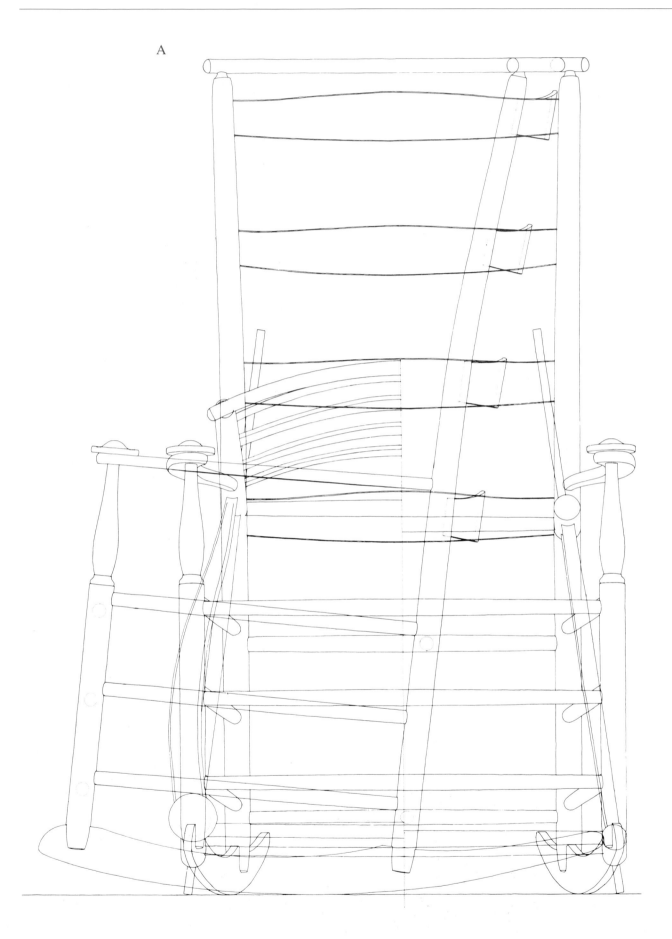

*Measured drawing by Erik Findsen, 1928. From the Furniture Collection of the Royal Academy's School of
Architecture, Copenhagen. Scale 1:5*

Four rocking chairs made by the Shaker sect of North America are shown. The measured drawing A and photographs B and C show three typical Shaker types. Photo D shows what could be described as a Shaker experiment. The design, craftsmanship and manufacturing of the Shaker sect are important elements in the historical development of industrial design. The Shakers have been described as the original Functionalists, predating, as they did, that movement in architecture by at least a century. They were active architects and designers and in everything they built and made there exists the beauty of union between function, form and technique. This attitude to design was part of their religion and it characterised their villages and farms, their houses and rooms. At the Philadelphia World Exhibition of 1876, in which the Shakers exhibited, they saw Thonet's steam bent chairs for the first time. This resulted in the design shown, D, in which steam bending has been incorporated; a good example of the Shakers' inventive talent.

A. *Early Shaker chair with back cushion-rail. Seat webbing not shown. Drawing from the Royal Academy's Furniture Collection, Copenhagen.*
B. *Chair of maple, similar to the drawn chair, at Kunstindustrimuseet, Copenhagen.*
C. *Chair of maple, New Lebanon, 1850.*
D. *Chair of hickory, Mount Lebanon, late 1870s.*

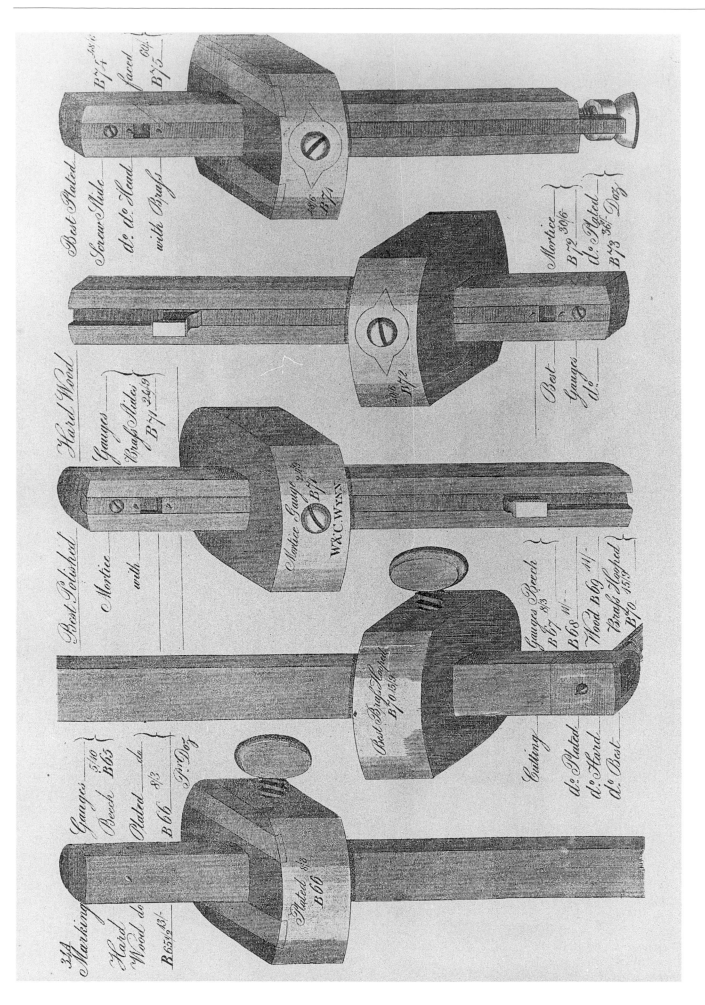

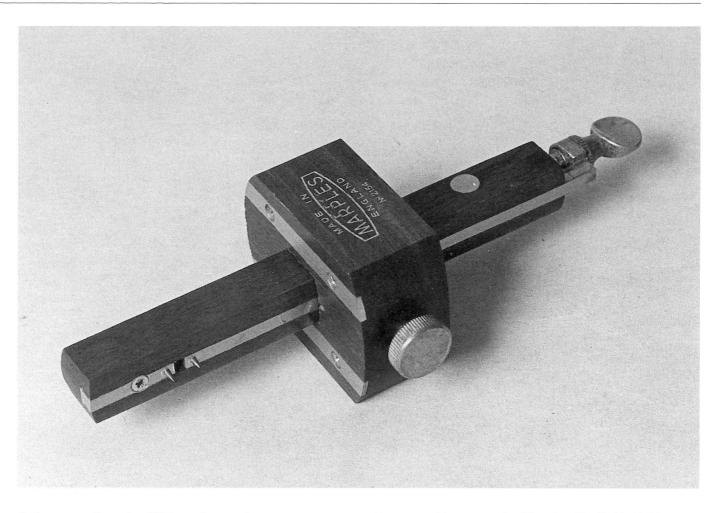

Left: a page from the 1892 catalogue of W.&G. Wynn, Birmingham, England.

Above: marking gauge by Marples, Sheffield, 1974.

Subject: The marking gauge is used by the woodworker for scribing guide-lines parallel to the face side, edge or end of the wood to which to plane or saw when thicknessing, jointing etc.

Manufacturer: The engravings are from the 1892 catalogue of the firm W. & G. Wynn of Birmingham. The photographed example is by Marples of Sheffield, 1974.

Design: Vernacular.

Materials: The photographed example is made of padauk wood and brass with steel spurs and screws.

Manufacturing processes: Skilled precision wood and metalworking with special machines. Final sanding of metal parts flush with the wood.

Dimensions: Overall length of the stock 19.5 cm.

Location: Author's collection.

References: *Dictionary of Tools* used in the woodworking and allied trades, *c.*1700 – 1970, by R.A. Salaman, London 1975.

Related products: Hand Tools, pp 138-139.

Evaluation: Both the top illustration on the left, and the photograph above show mortise gauges. Both are shown with the brass slide screwed a little out so that the adjustable spur is about 10 mm from the fixed spur. In this position the gauge could be used to mark out for a 10 mm wide mortise and a 10 mm thick tenon for example. On the reverse side of the gauge shown in the photograph is a fixed spur so that it can also be used for ordinary marking with a single line as a guide for sawing or planing. The two brass strips let into the face of the fence are simply to reduce the effect of wear. The remarkable thing about these illustrations (to the left from 1892 and above from 1974) is the similarity of the designs. The period of production is, in fact, even longer because the mortise gauge as shown in the catalogue was first manufactured in 1816. This is a good example of the way industrial precision sometimes helped to improve the quality of hand craftsmanship.

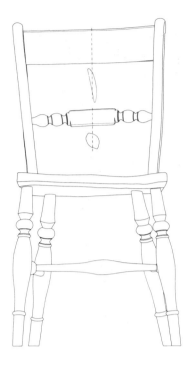
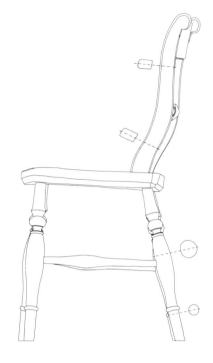

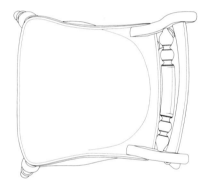

Measured drawing by Nis Øllgaard. Scale 1:10

Subject: Scroll back Windsor chair, so-called because of the scrolls which terminate the two back supports and because it comes from the same workshops that made Windsor chairs. The type was in production from the mid-1800s to the 1930s

Manufacturer: Unknown. Made in High Wycombe, England.

Design: Unknown

Materials: Seat elm, legs and stretchers of ash, the remainder beech. Stained yellowy brown.

Manufacturing processes: Legs and stretchers: bole of tree cut to lengths, cleft into billets, rough shaped with axe and draw-knife, turned on lathe. Seat: bole sawn to planks, cross cut, edge shaped on band saw, dishing formed by adzing and scraping. Back: supports shaped on band saw and edge profiled on front edge, top rail band sawn and profiled, lumbar support made on lathe or copy lathe. Assembly with animal glue.

Dimensions: The thickness of the seat and the depth of its dishing are a clue as to the age of a Windsor chair. If these dimensions are generous, as here, the chair probably antedates full machine production. Thickness 38 mm, dish depth 25 mm.

Location: High Wycombe Chair Museum.

References & notes: *The History of Chair Making in High Wycombe* by L.J. Mayes, London 1960. *The Windsor Chair* by Ivan Sparkes, Bourne End 1975. *Windsor Chairs* by F. Gordon Roe, London 1953. Article in *Arkitekten* No.11 by the authors, Copenhagen 1982.

Related products: Armchair pp 150-151.

Evaluation: When looking at this chair, it is important to remember that the High Wycombe chair industry started in the beechwoods of the surrounding Chiltern Hills. Judging by their construction, it is quite possible that the very first Windsor chairs were an invention of the woodsman. The original Windsors owe their very design to the

solution of the problem: how to make a chair with green material straight from the woods. In their green state most wood varieties are much easier to work but are subject to shrinkage and warping while drying. The woodsman chairmaker found the different trees whose wood was best suited to the various processes, which were performed by hand with the woods at a workshop. An alliance developed between these woodsmen and workshops in the local villages which led to specialisation: the woodsman made the chair parts and the workshops assembled them to finished chairs. This situation developed and the workshops became small factories, mostly in High Wycombe. One of the reasons for the success of the woodsmen, apart from their low prices, was that the legs, stretchers and spindles which they produced were from cleft wood. This meant that the grain fibres followed the direction of the finished part which therefore had optimum strength. In sawn parts there is

often the possibility of grain lying accross the direction in which strength is required. Fibre continuity was of fundamental importance for the thinly dimensioned parts and, when ordering, some government departments specified that legs and spindles should be of cleft wood.

The chair shown here is clearly a departure from the original Windsor and signifies, in its form and detailing, the industrialisation of Wycombe chair-making. There is not so much work for the woodsman, and the strength-giving arrangement of an arc of back spindles drilled into the seat, as in the classic Windsor, is here substituted by two uprights tenoned into the seat. The tenon, which is also used for the top rail joints, is a workshop joint which is foreign to the drill-jointed chair. So while this chair lacks the design wisdom of the proper Windsor, its compactness has made it useful in many situations where the Windsor was too big. This, and its low price, led to it becoming a 'standard' chair in Britain and abroad.

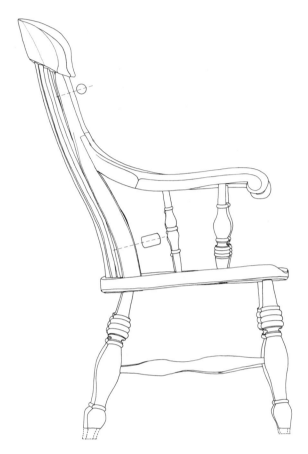

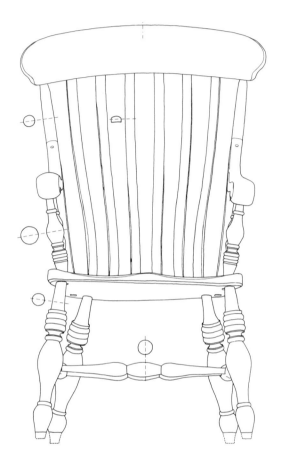

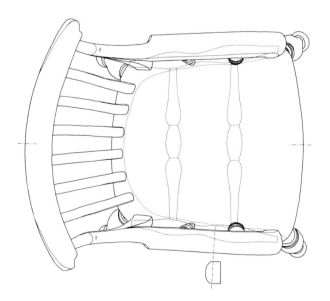

Measured drawing by Akis Nakagawa. Scale 1:10

Subject: Lath back Windsor chair. The name comes from the lath, or flat, section of the six back members. This was something of an innovation because they spread the load on the back of the seated person better than the traditional round sectioned spindles.

Manufacturer: Not known but probably one of the larger High Wycombe chairmakers. The type was made between the late 19th and early 20th centuries in very large numbers and is typical of the early factory Windsors.

Materials: Legs, stretchers, back supports, back laths, arm balusters: beechwood. Seat: elm. Head support and arms: cherry.

Manufacturing processes: This chair is late enough for many of the parts to have been machine made. The head rest, back supports, laths and arms are sufficiently uniform to have been band sawn and then spindle moulded to templates. The seat is too well formed to be machine dished and would have been hand adzed, shaved and scraped in the traditional way, after being cross-sawn from the plank, and band sawn around the edges. Legs, stretchers and balusters are very finely turned on the lathe, though probably not in the woods. As is typical for the Wycombe chair industry, unseen undersides are unfinished.

Location: Authors' collection.

The use of several wood types in the same chair was common. The whole forest was utilised. The flush joint, above, is therefore unusual as, apart from being costlier, it involved two different woods meeting flush, in this case: cherry and beech. This mixture of different colours and grains was concealed under a dark stain which often faded revealing beautiful pieces of wood.

References & notes: See p.148. Roundness of turned parts indicates whether or not they have been turned in the woods from green wood: if they are slightly oval in section they have been turned before drying. This chair is the only Windsor in the book with (almost) perfectly round sectioned legs and stretchers. The drawing shows the legs to have been shortened, a little more at the back than the front, to enable more relaxed sitting.

Related products: Comb-back Windsor pp 126-127.

Evaluation: The design of this chair is the result of an eotechnic industry's attempt to adapt itself to the Victorian taste and to the mechanised factory. However, the designer knew what he was doing: the Windsor concept of all components meeting at the seat is maintained; he utilises to the full one of the basic skills of the Windsor chair, turnery, to satisfy the opulent taste of the period; use of the spindle moulder enabled the profiling and shaping of the laths to give lumbar support to the back, which the stick Windsor had always lacked. Surprisingly, the designer used flush joints between the arms and the back supports – a joint belonging to fine London chairmaking, which was quite foreign to the characteristic bored joint of Windsor construction.

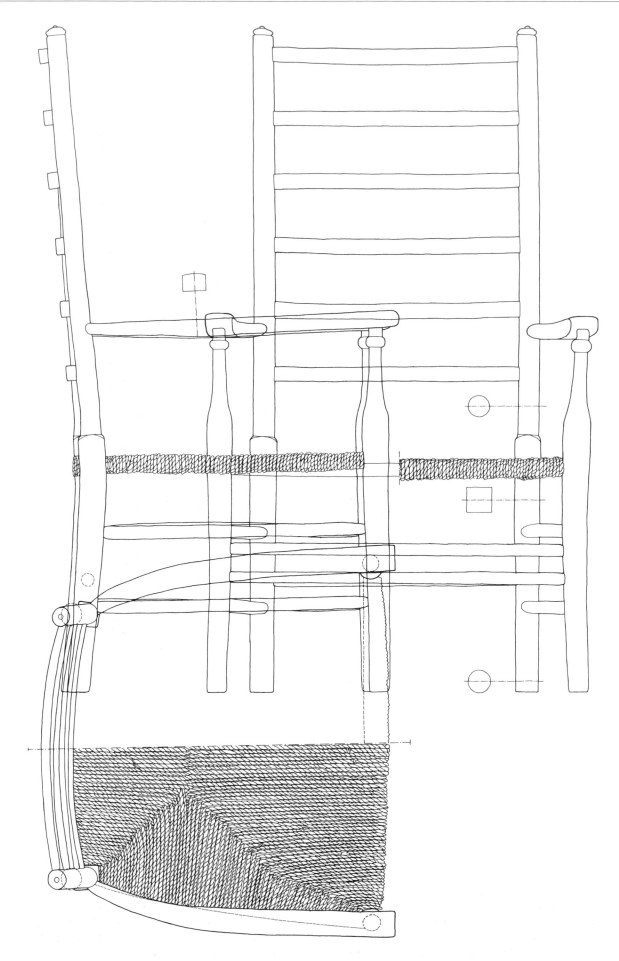

Measured sketch by Kirsten Broch Knudsen. Scale 1:5

It is sad that literature on the history of furniture so consistently ignores the simple vernacular. This little chair owes its many design qualities to the natural and uncomplicated placing and jointing of its 23 finely dimensioned parts into a useful, strong chair weighing only 3.5 kg. It was found outside a secondhand shop in London in 1950. Its design is in essence timeless though its stretchers of ramin – a wood imported from S.E.Asia – and its straightforward delicacy could place it a little later in the 1920s. The chair is quickly and cheaply made entirely with shoulderless joints. It is probably from the chairmaking town of High Wycombe. The legs and arms are of beech. The seat was originally rush. So simply can a beautiful chair be made!

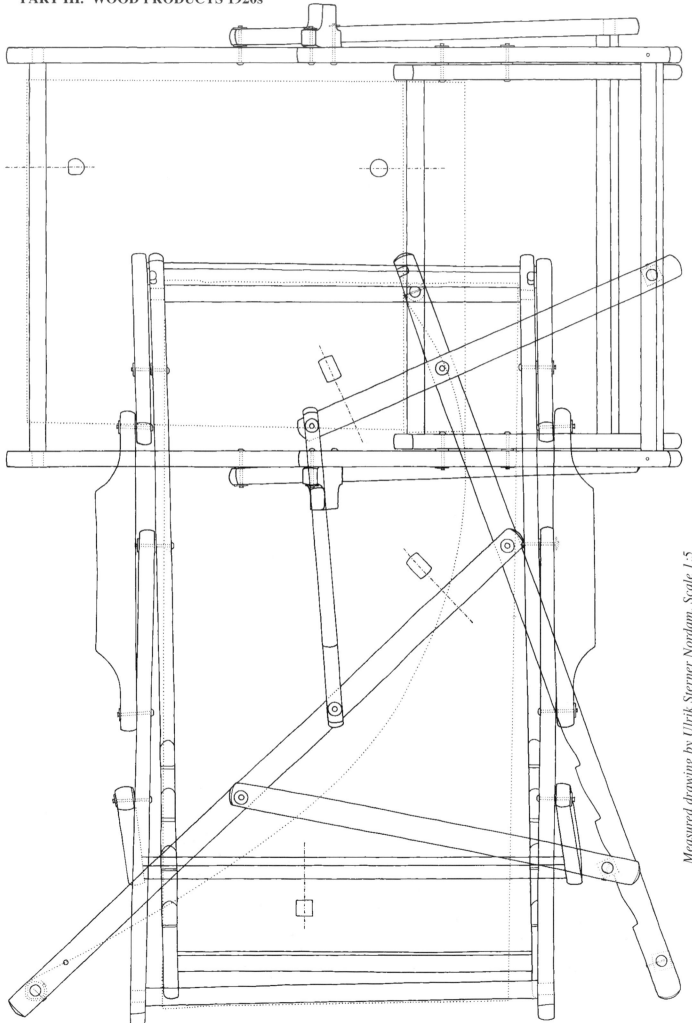

Measured drawing by Ulrik Sterner Nordam. Scale 1:5

Subject: Folding garden chair. A foot rest and a sun shade belong to the chair but are not shown.
Materials: Beechwood and canvas. Mild steel rivets, washers and nails.
Manufacturing processes: Rip sawing, thicknessing, moulding (3 profiles), cross cutting, turning (tenons), end rounding on band sander, notching on bandsaw, boring. Assembly with round tenons cross nailed, and rivets. Canvas hemmed and fastened with tacks.
Dimensions: Standard section of principal members 22 x 32 mm, weight of chair alone 4.2 kg.
Evaluation: This is an anonymous chair type which has been made in many countries and in countless versions. Many of these are of better quality and of heavier

dimensions, but the one shown here fascinates us because it is extremely simple, is made with the utmost economy of materials, is quickly – but intelligently – made and has proved remarkably robust. This example has been used nearly every summer since the 1930s, albeit with several changes of canvas. Most important, and unlike many related versions, this chair has arms which are almost a necessity when sitting in reclining positions. This type of chair in which one's body is bent to the sag-line of the canvas is not good for people with weak backs unless a cushion is placed in the lumbar region. It can be seen from the drawing that the chair has become 'loose' and a certain amount of warping has occurred but this does not make it a less loved summer chair.

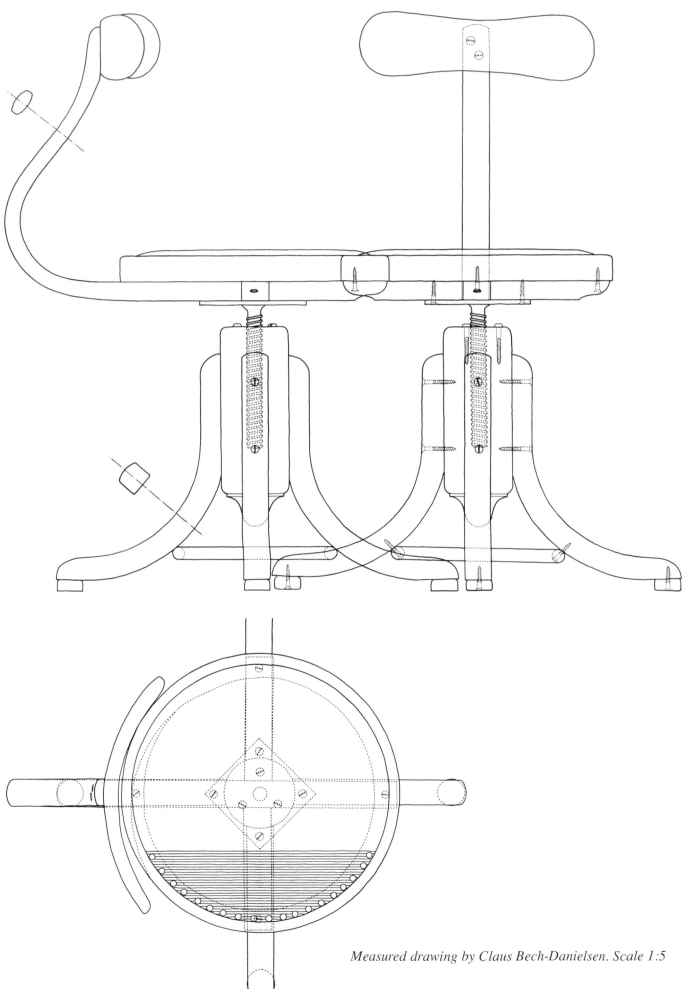

Measured drawing by Claus Bech-Danielsen. Scale 1:5

Subject: Office chair with adjustable seat height.
Manufacturer: Fritz Hansen, Denmark.
Designer: Unknown, probably the manufacturer.
Materials: Beechwood, steel height-adjustment screw welded to fixing plate under seat and turning in 50 mm deep threaded iron casting screwed to top of central wood cylinder. Seat of corduroy over thin padding on plywood.
Manufacturing processes: The chair consists of eight steam bent parts and a central turned cylinder all individually machined, finished and assembled solely with screws. The cylinder, a cross rail under the seat and the feet studs are the only wooden parts which are not steam bent.
Dimensions: Seat, max. height. 57 cm, min. height, 43 cm.
References & notes: We first saw this chair at an exhibition at Louisiana Museum (Denmark) in 1974. The exhibition catalogue is part of the Museum's periodical *Louisiana Revy* 14th volume, no. 3-4, 1974.
Evaluation: In its construction this chair belongs to a family of chairs founded by Thonet in the 1870s and later followed by versions in steel tube and rod. Seats are supported either on a spindle or, as here, on a height

adjusting screw encased in a column to which the legs are fastened. To the revolving seat could be fixed a variety of back rests sometimes with arms. Fritz Hansen's own catalogue from the 1940s shows variations on the theme in wood and in a combination of wood and metal. But this chair has an appeal and a charm which no later office chairs had. It combines well-shaped and machined components, and their assembly, to an articulated whole. It embodies the maximum of adjustment and mobility that can be achieved in a wooden chair. From this time onwards the office chair became a contraption which could adjust in nearly every direction, another machine along with the typewriter and calculator to help in the ever increasing rationalisation of the office. Like the Thonet chairs, it demonstrates the basic principle of quantity production: finished parts simply assembled to form the finished product. The natural light beech enhances the nicely shaped parts. The well-formed tiny lumbar support on its springy stalk would be excellent if it were 25 mm lower at lumbar height and did not pinch one's back ribs. This was a sympathetic Danish contribution to the office interior at the height of the period of hard functionalism.

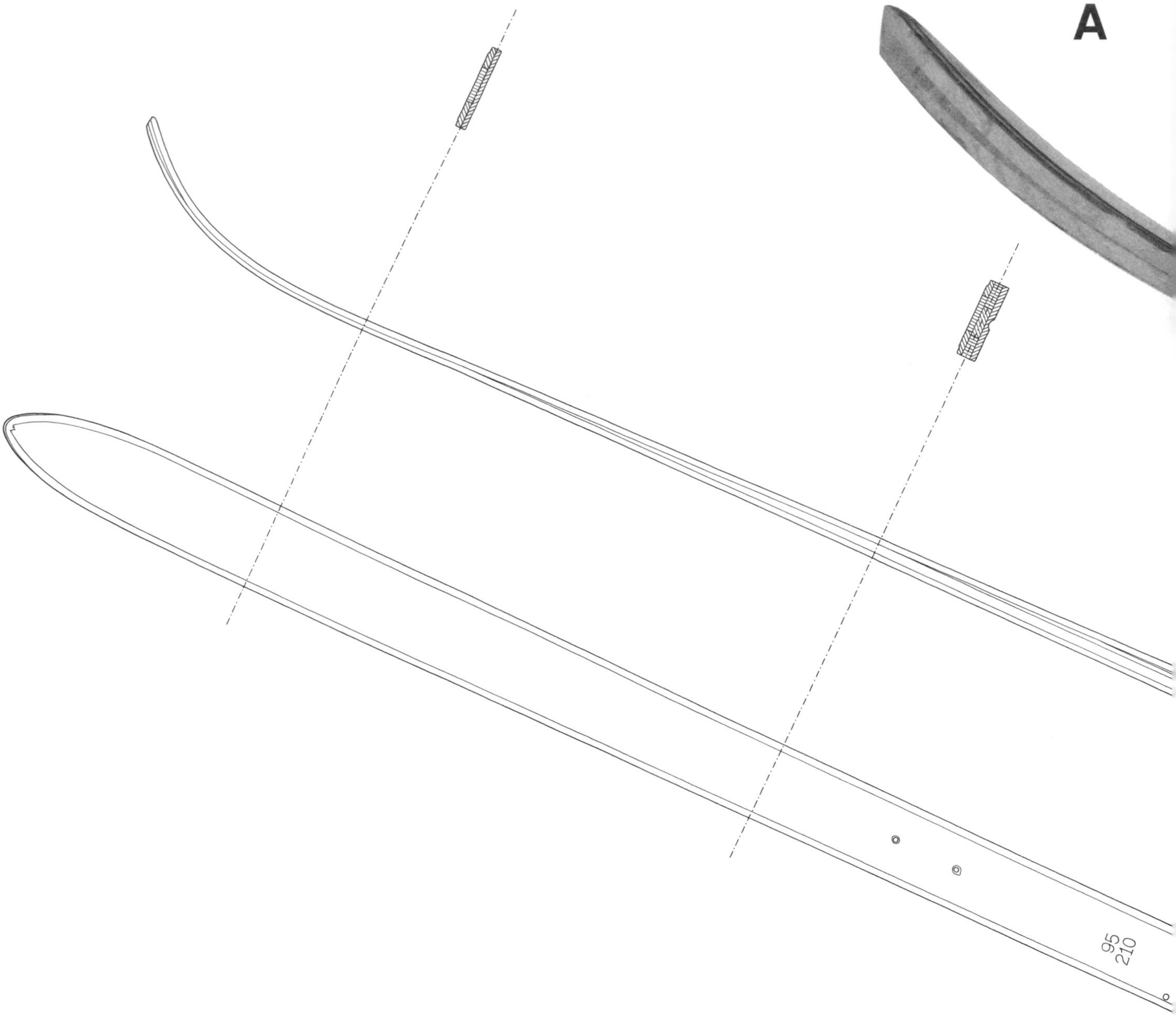

A

95
210

The ski. Some people – including the authors – are fascinated by the design and construction of certain types of sports equipment. This is because here materials are employed in a specially interesting way. The racket, bat or stick has to perform with optimal efficiency usually requiring great strength with lightness, resilience, balance and handleability and this combination gives them a functional beauty. Behind their design, too, lies decades of user and maker experience, so they often embody the techniques of past and present. The origins of some of this equipment are to be found as the tools of daily life where function and reliability were a necessity. The ski is one of these and it teaches us a great deal about the potential of constructions in wood. We have made a measured drawing of a perfectly ordinary factory-produced ski, probably from Norway, which historically represents the final stage of wooden ski design before plastics started to take over. It is built up of 9 strips of

well dried, straight grained ash of various dimensions which in production are alternately laminated (glued together under pressure) and moulded in such a way as to stabilise the wood and resist its natural tendency to twist. The strips are laid and laminated with the annual rings turned in different directions, as shown in the sections, so that the forces in the wood work against each other and are neutralised. This is fundamental because even a slightly twisted ski will not run straight in the snow. The front and rear ends are reinforced by cross grained wedges which bind the main strips and prevent the tendency to split due to impact or water absorption. See details A and B. The groove along the middle of the sole is to minimise side-slipping in the snow. In use, the ski is subjected to considerable loading – at one moment as a bridge, at the next as a cantilever, and the next as a spring and shock absorber. All this is expressed in its beautiful form and construction.

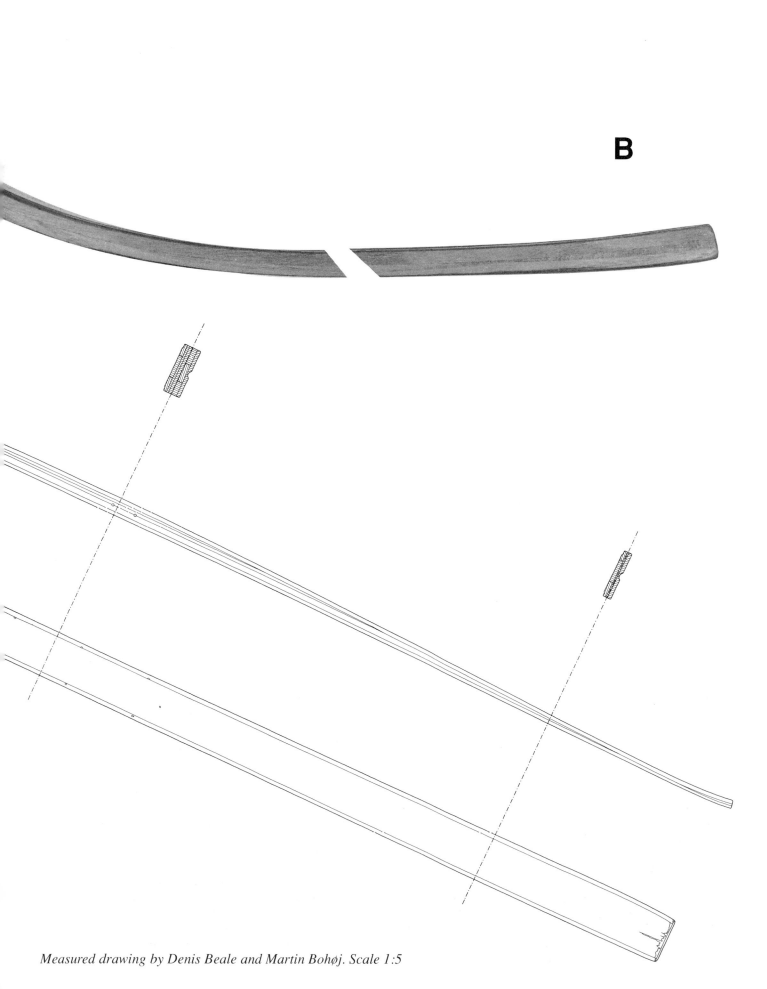

B

Measured drawing by Denis Beale and Martin Bohøj. Scale 1:5

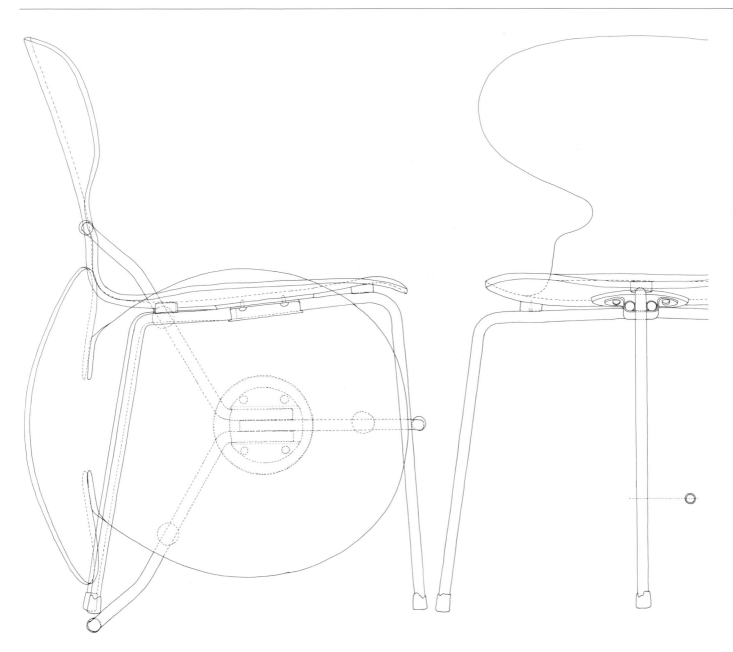

Measured drawing by Poul Slivsgaard Køster. Scale 1:5

Subject: Chair for auditoriums and general use in public buildings. A later model has four legs.
Manufacturer: Fritz Hansen A/S, Denmark.
Designer: Arne Jacobsen.
Materials and manufacturing: Seat and back shell: 10 veneers, band-sawn to size, glue applied and pressure formed in hot moulds. Edge machining to final shape, disc for leg fixing glued on, sanding, lacquering. Leg unit: 14 mm x 2 mm steel tube, nickel plated, welded together. Assembly: 3 rubber blocks screwed on under seat, leg unit fastened to shell with screwed-on 'U' plate under the seat. Plastic feet fixed.
Dimensions: Shell 9 mm thick. Weight 3.5 kg.
Location: Author's collection.
References & notes: This description applies to the chair shown here which is one of the very first examples.
Made In Denmark by Arne Karlsen and Anker Tiedemann, Copenhagen 1960.

Dansk Møbel Kunst vol. 2, by Arne Karlsen, Copenhagen 1992.
Danish Chairs by Nanna & Jørgen Ditzel, Copenhagen 1954.
Jacobsen by C. Thau and K. Vindum, Copenhagen, 1998
Evaluation: This little chair was something of a design event when it appeared. It is still an astonishing chair over 40 years later, and an established element in many buildings. How is it that this advanced design has become a classic in its own age? The 'Ant', as it is called in Scandinavia, comes from the hand of an architect who designed most of the furniture and fittings for his buildings. They became elements of his architecture and this often gave them their own independent strength. The chair functions exceptionally well: it is flexibly comfortable, it is light and easy to handle, it is minimal and yet tough, it stacks easily and closely. It is both sculpture and public servant.

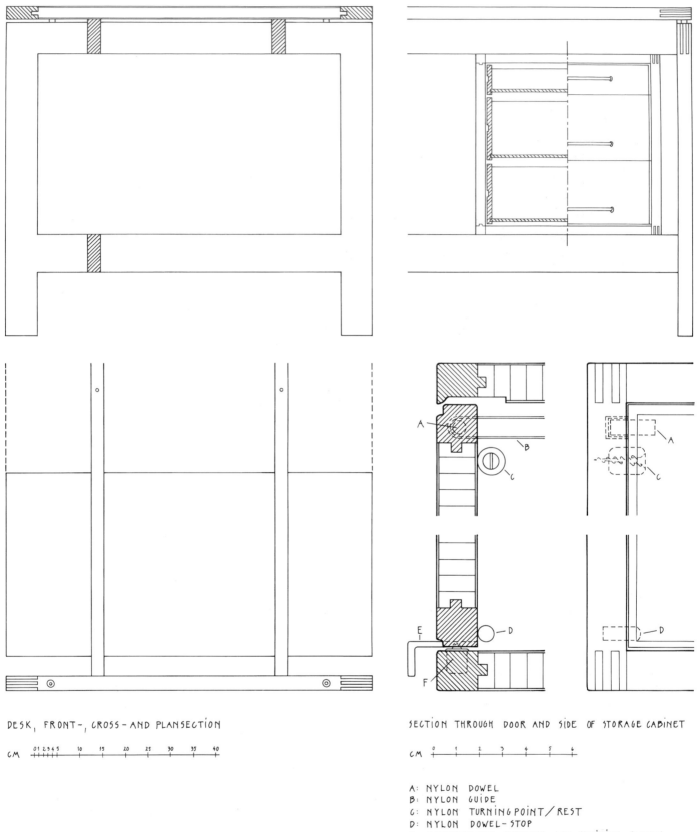

DESK, FRONT-, CROSS-AND PLANSECTION

CM

SECTION THROUGH DOOR AND SIDE OF STORAGE CABINET

CM

A: NYLON DOWEL
B: NYLON GUIDE
C: NYLON TURNING POINT/REST
D: NYLON DOWEL-STOP
E: MATCHROMED HANDLES AND STRIKING PLATES
   FOR BALL LATCH
F: BALL LATCH

Measured drawing by Poul B. Eskildsen.

Photo: K. Helmer-Petersen

Subject: M40 modular office furniture series consisting of conference tables, typewriter tables, shelf, cupboard and drawer storage.
Manufacturer: Munch Møbler, Denmark.
Designers: Henning Jensen and Torben Valeur.
Materials: Oak wood. Structural members and edge frames, lippings and drawers of solid oak; tops and panels of oak veneer on blockboard. All fittings are matt chromium-plated metal.
Manufacturing processes: Structural and frame members: bole of tree sawn to planks, drying in kiln. Rip sawing, thicknessing, moulding to required section, size and profile, cross sawing to required lengths, drilling for bolts, fittings and dowels. Sanding, finishing. Panels, doors, and work-tops: preparation of material as for structural parts. Veneer knife-sliced from tree bole. Veneer leaves edge-joined and cut to approximate sizes required. Blockboard sawn to panels and tops of approximate sizes required. Veneer glued and pressed to these on both faces. Edge-framed and lipped with glue in presses. Moulding and drilling, sanding, finishing. Drawers: moulded and jointed in solid oak with oak-veneered plywood bottoms. Finally: assembly or flat packing. Many of these processes are conducted in sequence and with several units at once, on multi-function machines.
Dimensions: Based on 40 x 40 cm module in plan.
Location: Widely used in Scandinavia in both large and small offices where quality is appreciated.

References & notes: The series is still in production at the time of writing, having been augmented through the years.
Munch Møbler catalogues from 1961 onwards.
Mobilia magazine, no. 69, April 1961.
Dansk Kunsthåndværk, no. 8, 1962.
Evaluation: (based on the designers' description): Designed on the basis of a 40 cm module the M40 office furniture offers a maximum choice of combinations. The basic principle of the system is that all horizontal dimensions must be multiples of this module. Vertical dimensions are decided by user-functions and sizes, and conform to mutually defined height levels. All storage units are multiples of 40 cm above the base. Behind the adoption of such a rigid system of fixed dimensions is a desire to give harmony to small groups as well as larger arrangements of units. Work-top framing, with its expressive finger-jointing at the corners, and the legs, are of the same breadth and thickness which 'explains' the construction and gives a feeling of quiet, controlled simplicity. To facilitate packing, transport and storage, the larger units are assembled with 'unbrako' bolts. Drawers for hanging folders are equipped to take various sizes of folder. Doors swing up and in, so that they remain open but out of the way during office hours. The surface finish is a lacquer type which protects the wood but allows it to remain natural in appearance, as if it were newly planed.

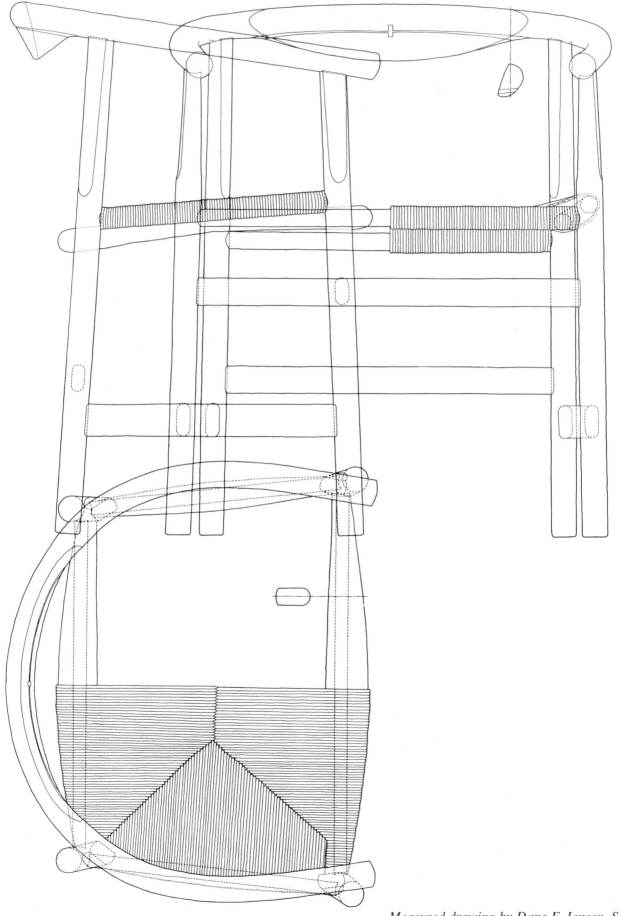

Measured drawing by Dane F. Jensen. Scale 1:5

Subject: General purpose bow chair.
Manufacturer: PP Møbler, Denmark.
Designer: Hans J. Wegner.
Materials: Sycamore with glue-fillet of wengé wood in the back rest. Seat frame of beech. Seat of paper cord.
Manufacturing processes: Back-and-arm bow form pressed of 3.5 mm laminates, lumbar support turned on copying lathe, jointed, formed on spindle moulder. Legs turned, facetted on spindle moulder. Stretchers planed and moulded. Mortice and tenon joints throughout. Components sanded. Assembly. Second sanding. Soap finishing treatment.
Dimensions: Height 71 cm, seat height 44 cm, width 58.5 cm, depth 49.5 cm.
Location: Author's collection.
References : *Hans J. Wegner, en stolemager,* Danish Design Center, Copenhagen 1989.
Dansk Møbel Kunst i det 20th. århundrede, vol. 2, by Arne Karlsen, 1991.

Evaluation: The construction of the bow chair is exceptional in that it surrounds the seated person with back and arms and uses the bow to give strength. This is about the eighth bow chair which Wegner has designed. It is a theme which he has taken up and renewed at intervals ever since his first 'Chinese Chair' of the early 1940s. We have chosen it from Wegner's immense production because it embodies many of his qualities as designer, architect and chairmaker. It also epitomises the best of the latest remarkable 50 years of Danish furniture making. The chair is light and yet robust, good to hold and handle, is fresh and new and yet has its form and technique well planted in history. Apart from his eye for structure and space Wegner uses the machine tool to both define and relate the different members of the chair. Together these visual and craftsmanlike skills give his chairs a special place in the history of industrial design.

Subject: Mixing spoon and spatula.
Manufacturer: Scanwood, Ryslinge Trævarer, Denmark. Founded 1919.
Design: Traditional.
Materials: Spoon, sycamore. Spatula, beech. The wood is machined while still green (undried).
Manufacturing processes: Boles of trees converted to quarter sawn planks, cross cutting to blocks.
The spoon: 1 Scooping two bowls with spindle knife moulder.
2 Separating the two potential spoons by band sawing.
3 Shaping the profile of the bowl and handle on spindle moulder. **4** Rounding the handles on a spindle rounder (resembling a multi-cuttered pencil sharpener) **5** Shaping the under side of the bowl on the spindle moulder. Drying. Surface finish acquired by tumbling in a rotating drum tumbler. Sorting.
The spatula: Repeated band sawing of the blocks to the 3 mm thick curved sweep of the spatula (on the quarter sawn face). Shaping the edge profile to form handle and blade with spindle moulder. Drying, tumbling, sanding, and disc sanding to form chisel blade end. Sorting.
Dimensions: Spoons: eight sizes are made in sycamore between 87 cm and 27 cm in length.
Spatula: 29 cm long.
Location: Author's collection.
References: The manufacturer.
Evaluation: When these simple tools were suggested to represent wood products of the last two decades we shied at the idea. Here were no innovations, on the contary. But the more we thought about it and the closer we looked at them and their industry the more interesting it became. Although the wooden kitchen tool industry has led a humble existence and is of ancient origin it is not a bad idea, in the midst of the present flood of commercially steered technology, to remind ourselves that some materials and processes – and the greater part of personal requirements – have remained unchanged throughout the last hundred years. Wooden tools, the whole category from brushes to axe handles, are still the best to hold and use.

Wood is the only material in the book which is 'ready made' from nature's hand. After it has been sawn to manageable sizes and dried, we can form it in various ways, as we have seen, to make it into components of larger products. This is not the case here. Each finished product is made by the multiple machining of one piece of wood. The machine tools are designed and made especially to perform the various shaping operations. The process is exceptional in that while it is normally important for wood to be dried before machining, to establish dimensional stability, here it is machined in its wet state and dried afterwards. This is because the special machine cutters can cut the wet wood easier and give a cleaner cut (see Copying Lathe, pp 38-39, where the same applies). It also allows quicker and more economical drying since there is a smaller volume of wood to dry. When the piece is dried it shrinks across the grain. This change of dimension is in this case unimportant since the product is independent – it is not required to be assembled with other parts. Because the material is quarter sawn the twisting and distortion during drying are kept to a minimum.

The wood chapter started at the beginning of the 18th century with a wooden product, the trug, which at that time was already ancient. It ends, after 300 years, with another timeless product. The material itself has, in each case, inspired man and shown him the way to industrial design solutions to two very different problems.

Stages in machining the wooden mixing spoon.

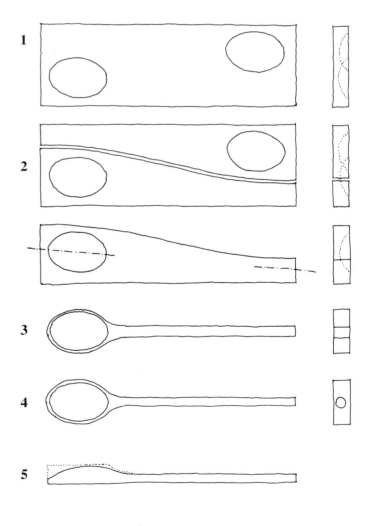

The spatula and spoon are chosen from a large number of other wooden cooking tools – the variety of types and sizes made is reminiscent of the tool and hardware catalogues of the last century (see pp 88-89 and 96-97).

PART IV

CERAMIC PRODUCTS

'**K**eramos' is Greek for 'potter's clay'. This name was given to one of the world's oldest industries because clay possesses a number of unique properties. The three most important are:

1. The material is found throughout the world as a sediment of granite and other rocks. Clay is easily extracted because it is often deposited by water, and is therefore found in large quantities close to the earth's surface.
2. In a moist condition the material is plastic, so that objects can be formed with it, either as impressions or as self-supporting hollow or solid shapes, which after drying become permanent.
3. The material is resistant to high temperatures, and is unaffected by water after it has been fired. During firing clay undergoes a chemical change which binds the clay crystals together and gives the material unsurpassed wearing qualities and long life in daily use.

These three qualities are characteristic for each of the product groups into which the ceramics industry can be divided: *heavy ceramics,* used mainly in building, such as bricks, roof tiles and floor tiles; *fine ceramics,* which covers domestic table ware and utensils, containers for foodstuffs, works of art, and ornaments; *technical ceramics*, including highly insulating products, as well as ordinary insulators, chemical resistant products, and hard-wearing components for many industries.

Within these main categories we find most of the classic production processes represented. Heavy ceramics are mainly produced by *pressing* and *extruding.* The clay is formed under high pressure in metal moulds, or forced out through dies, and cut into units of the required shape and length, after which the item is fired. This produces a solid clay product in which the fired clay's ability to withstand compression, exposure to the elements, and the wear of use, are exploited to the full. The fired clay is porous and, as opposed to concrete, able to regulate the climate in buildings by absorbing and releasing moisture.

Note: in the practice of ceramics the word 'throw' means to turn so as to shape a pot (or other object) on the potter's wheel. The word 'turn' means shave-taking so as to trim, or finish, an already thrown pot, often in leather hard consistency, as on a lathe.

Fine ceramics employs the classic processes: hand throwing, moulding, template turning, and casting. *Hand throwing* is probably the oldest ceramic production technique. It is performed without the use of tools, other than the human hands and the rotating potter's wheel. Centrifugal force is applied with a minimum of human energy to create hollow products which can be varied in shape and size according to what is required of them.

CERAMIC PRODUCTS

Hand moulding is the usual way to produce clay articles where the material is either too plastic or too coarse to be used in other processes. Production is carried out with simple moulds whose shape is negative of either the outside or the inside of the pot. The side which is not in the mould is formed by hand. The pot can be separated from the mould after drying.

In *template turning* the template for the inner or outer form is held and controlled by a holding arm; it is by way of being the mechanisation of the human arm and hand. This has led to the automation of the work with a precision in reproduction seldom seen in thrown pottery. Here we find the light ware such as cups, saucers, plates and bowls which have some of the qualities of thrown pots such as varying body thickness between bottom and side. But unlike purely thrown pots, special provision has to be made in mould design to allow 'undercut', or 'returning' shapes to be released from it.

The *casting* process is perhaps as old as throwing. This always employs a mould which contains the whole pot shape in negative. Moulds of fired clay, or plaster of Paris, are porous, so that they can absorb the water in the pourable, wet, clay mix. The clay (casting slip) adheres to the sides of the mould in varying thicknesses depending on the time it is left in the mould. Casting is best suited to porcelain and earthenware clays which are not too plastic in their composition, and which relatively easily relinquish water, and dry in the mould. These clay types, which are often compositions of different clays and minerals, are ideally suited to the industrialised ceramic manufacturing process, where there has to be careful control of the composition of materials and the uniformity of the end product.

Technical ceramics include a large number of different components, large and small, used in many industries primarily in its capacity as electrical insulator. The range is very extensive – and increasing – from spark plug parts to high tension insulators of several meters in length. Here pressing and turning are the predominant methods of production. The dust pressing technique is also frequently used because it gives much greater dimensional accuracy. Dry clay dust is moulded under high pressure in metal moulds, and due to its dryness, shrinks much less during firing than items of moist clay. The accuracy achieved by dust pressing enables ceramic components to be assembled with other types of component. Large insulators are turned by hand from solid, leather-hard porcelain cylinders. This can only be done by specialist craftsmen who can judge precisely the degree of dryness which is suited to the particular shape to be turned. In the chemical and pharmaceutical industries ceramics have been used for centuries as the favoured material for containers and equipment which have to be resistant to chemical attack. These containers, with their chemical resistant glazes, are some of the most eminent ever made within utility ceramics.

The principal ceramic groups are *earthenware, porcelain* and *stoneware*. Their names are derived from the compositions of clays employed, and from the firing temperatures used. Earthenware is a very broad group ranging from ware with white or cream coloured body, of which there are a number of examples shown here, to the soft, reddish brown ware, glazed or unglazed, on pp. 210-211. Here we have only shown one production diagram – that of porcelain – to indicate the sequence of processes which apply to the production of nearly all ceramics.

Snorre Læssøe Stephensen

The products on pp 178-185 were recorded with the co-operation of the Trustees of the Wedgwood Museum, Barlaston, Staffordshire, England.

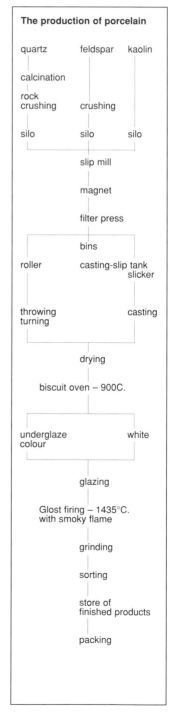

The production of porcelain

quartz   feldspar   kaolin

calcination

rock crushing   crushing

silo   silo   silo

slip mill

magnet

filter press

bins

roller   casting-slip tank   slicker

throwing turning   casting

drying

biscuit oven – 900C.

underglaze colour   white

glazing

Glost firing – 1435°C. with smoky flame

grinding

sorting

store of finished products

packing

170

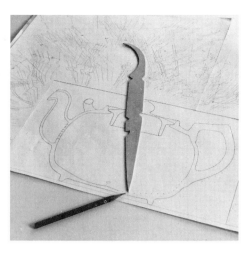

1   2

Fig.1 *Arch fixed at right angles to baseboard with drawing paper fixed to it.*
The subject (teapot) fixed with its centre line corresponding to the same plane as the drawing paper.
Feeler, made with a distinctive form, for drawing to.
The form of the teapot's inner surface is transposed to the paper as a fan of feeler lines. The outer surface has already been transposed.

Fig.2 *The drawing paper is removed from the arch and laid horizontally. A piece of drawing paper is taped into the arch opening.*
The procedure is now reversed and the feeler lines are used to guide the feeler so that dots can be made where the point of the feeler had previously touched the teapot. When all the dots are placed they are carefully joined and the first pencil drawing of the teapot's exact section appears.

A similar procedure is used to establish the plan of the subject, though carried out horizontally.

The graphic recording of hollow ware

There are ways and means, many of them the draughtsman's own invention, of accurately measuring and drawing nearly all objects, but when we came to hollow ware, pottery and glass we were stuck – and nobody seemed to be able to help. We had to reject many good ideas, from the beautifully simple such as placing the subject with its axis at right angles to the sun's rays and drawing round the shadow, to the more advanced such as using a laser. Even the photogrammetric computer method, as we knew it, could not help us. None of these solutions were adequate because in hollow objects one of the most important measurements is the most illusive: *wall thickness*. Since, in a pot or a glass object, wall thickness varies from one part to the other, we needed more than just thickness measurements. We needed them in their correct positions. We needed to be able to draw the inside surface line and the outside surface line in their correct relationship to each other.

Two students hit on a very good idea. They placed the subject exactly half way under an arch cut out of a piece of stiff sheet material, such as 6 mm plywood, fixed at right angles to a horizontal base panel. With the help of graph paper stuck to the surrounding wall of the arch and specially shaped, thin plywood feelers with pointed ends they could plot the many points required from the inner and then the outer surfaces. This was done by measuring vertically and horizontally and transferring the measurements in the form of dots to another piece of paper. The required lines were then drawn by connecting the dots. The ewer on p. 196 is drawn by this method.

The chief disadvantages of the method were its slowness and its dependence on rule measuring, which is insufficiently accurate for smaller manufactured items. Martin Bohøj found it necessary to abandon measuring altogether. He combined the arch arrangement of the method explained above with a technique used by flooring experts and other building crafts when, for example, carpeting has to be scribed to a complicated run of skirting. They do this by establishing the salient points on a piece of paper lying on the floor opposite the skirting with a feeler, or probing lath. A template of the skirting can then be cut in the paper and the carpet cut to this scribe-line.

In doing a measured drawing of a hollow, free formed object a similar technique can be used – though, to start with, on a vertical plane. The photographs describe the method which has been used to measure most of the items in the sections on products of ceramics and glass.

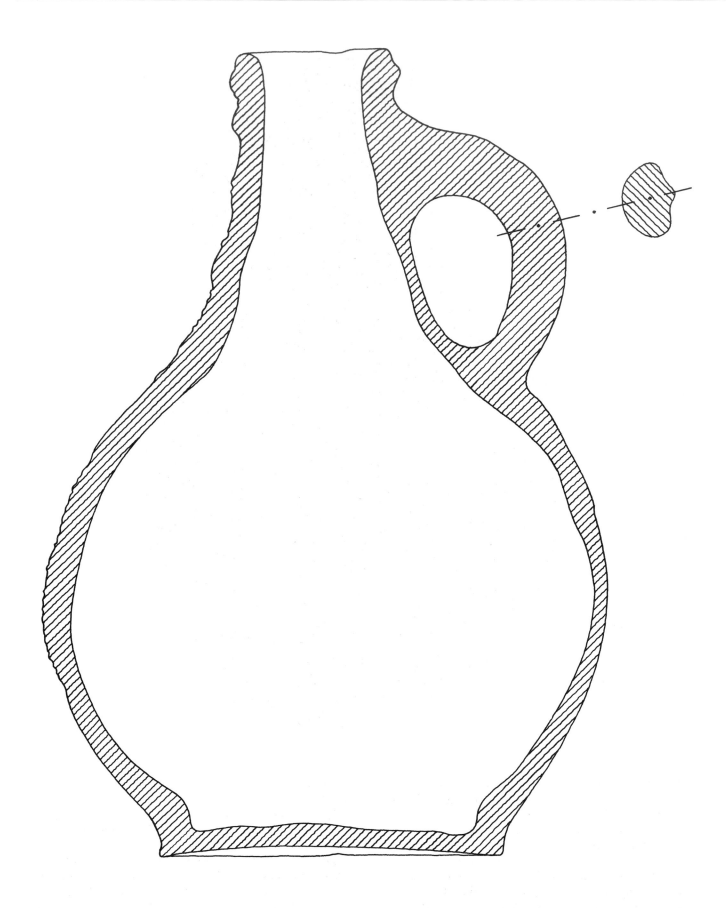

Measured drawing by Martin Bohøj. Scale 1:1

Subject: Bellarmine beer jug, late 17th century.
Manufacturer: Unknown, probably English. Originated in Germany and Flanders, 16th century.
Materials: Salt-glazed stoneware.
Manufacturing processes: Hand thrown. Then turned when leather-hard, to get a smoother outer form and make the foot and neck profiles. The decorations are cast in separate pieces of the clay and joined to the corpus (sprigging). The handle is likewise sprigged to the corpus. During firing salt is thrown into the kiln when at 1200°C. The salt decomposes to sodium and chlorine gas. The sodium combines with the quartz in the clay to form a thin glaze.

Evaluation: The name Bellarmine jug is derived from the Flemish and German 16th and 17th century tradition of making beer jugs with a bearded face on the neck. The jugs were given the name Bellarmine when the face became a caricature of the Jesuit cardinal of this name whose preachings had earned him the ridicule of the Flemish Protestants. The jugs were exported to England and eventually production was started there. In the present context the importance of the jugs is in the salt glazing of stoneware. This was a German invention of the period whose physical strength and resistance to corrosion are still amongst the best in ceramics.

173

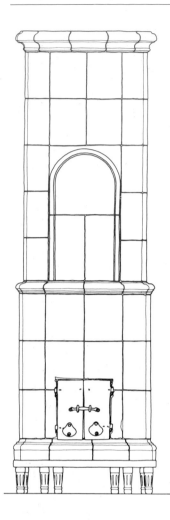

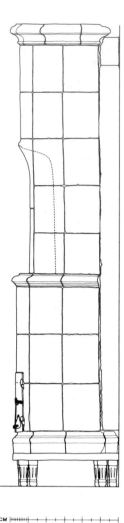

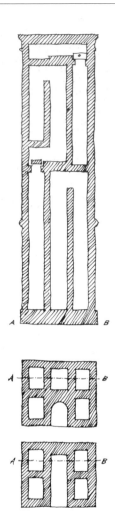

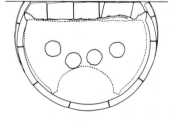

CM |ⱭⱭⱭⱭ| + + + + + + + +
0   10

Cross and plan sections showing one of Cronstedt's recommendations.

Measured drawing by Claus Beck-Danielsen.

Subject: Swedish kakelugn (tiled stove) for room heating with wood fuel.

Manufacture: Mariebjerg Fajancefabrik, Sweden.

Design: Made according to design principles laid down in 1775 for fuel economy by C.J.Cronstedt.

Materials: Externally, hand decorated tin-glazed earthenware tiles. Iron leg frame and fire-chamber doors. Internally, fireclay duct bricks.

Manufacturing process: The smooth and the profiled tiles are pressed in moulds. After drying and first firing, the tiles are glaze fired, with decoration. The fireclay bricks are pressed in special moulds. On site: The bricks are built up with mortar joints to form a ducted structure, and then tiled.

Dimensions: Height 258 cm, greatest diameter 80 cm.

Location: Ordrupgaard Museum, Copenhagen.

References & notes: *Kakkelovn og Jernovn,* by Ebbe Johannsen, Copenhagen 1980. *Kakelugnar,* by Britt och Ingemar Tunander, Västerås 1973.

Related products: Solid Fuel Stove, pp 106-107.

Evaluation: This type of stove was highly developed in northern Europe, and at this period they were quite sophisticated examples of industrial design. Though they appear simple they are both ingenious and quite complicated inside. They work on the principle of the storage heater: the gradual heating up of a large mass of brickwork which in turn gives off warmth over a long period. This is achieved with remarkable fuel economy, by leading the hot smoke through a labyrinth of ducts on its long way out to the chimney. With the help of careful draught control and small logs, closely laid, a high temperature is obtained in 3-4 hours. The stove then continues to radiate a pleasant warmth long after the fire has gone out. 'Kakelugnar' still stand like great silent figures in the rooms of older Scandinavian houses, permanent reminders of the long cold winters.

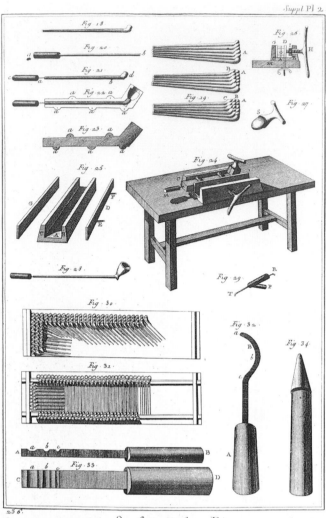

Art de faire les Pipes

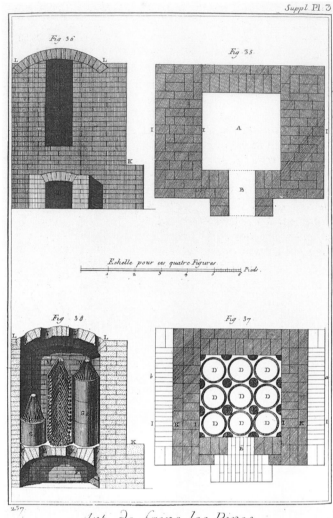

Art de faire les Pipes

Throughout the first 200 years of the import of tobacco to Europe (begun in the 16th century), the clay pipe was the smoker's friend. Pipe making became a speciality amongst the potters of England and Holland, and quickly developed into a clay pipe industry, capable of supplying the market with many different types. The clay pipe was developed from the wooden pipe of the American Indians, from whom the custom of smoking came. Like its wooden predecessor the clay pipe insulated the heat of the glowing tobacco, and had the additional advantage of not burning. The fired clay is brittle and the thin pipe is sensitive to impact, so pipes broke frequently in use and during transport. Contemporary accounts put the pipe smoker's consumption at 20 to 40 pipes per year! The first 'use and scrap' industry was born.

The pipe in the photograph is made of plastic, white-firing clay imported from Holland. This, so called, 'pipe clay' consists of very fine particles which give the clay strength during drying, and after firing. The plastic clay is rolled out in thin layers, with a thick portion at one end, and made into elongated rolls. A roll is laid in an oiled mould of brass whose two parts each contain half of the pipe's shape in negative. The moulds have locating pins which make the two halves fit precisely together.

Before the moulds are pressed together, a thin rod tool is pushed through the centre of the roll to form the smoke tube. The brass mould is open at each end so that, when pressed, excess clay is squeezed out and removed. The hollowing of the bowl is done with an oiled, steel cone-shaped tool, which is carefully pressed down into the pipe head until the point of the cone touches the steel rod and there is a clear passage through. The pipe is then removed from the mould and imperfections cut away or repaired. Finally the maker's name is pressed into the heal of the pipe.

At Rømer's factory the pipes were fired with wood fuel in high, cylindrical saggars. The firing temperature was between 900 and 1,000°C. The pipes were treated with wax to make the surface resistant to dirt and liquids. Instead of the wax treatment some of the pipes were glazed in cobalt blue, or copper green lead glaze and fired again.

The robust press mould, the refined hole making techniques, and the systematic firing have made it possible to produce clay pipes at a sensible price right up to the present day.

Snorre Læssøe Stephensen

This pipe was manufactured by Rømers Tobakspibefabrik of Nørresundby, Denmark.
The type shown was the so called 'bondepibe' (farmer's pipe). The smoke tube measures 20 cm from bowl to mouthpiece.
Most of the processes described and the equipment and tools used can be seen in the two engravings from Diderot's Encyclopaedia.
The pipe, which is one of the few (broken) survivors, was photographed at 'Pibemagerhuset' (Pipe maker Rømer's house) which is part of the Sundby Collections (museum), Nørresundby, Denmark.

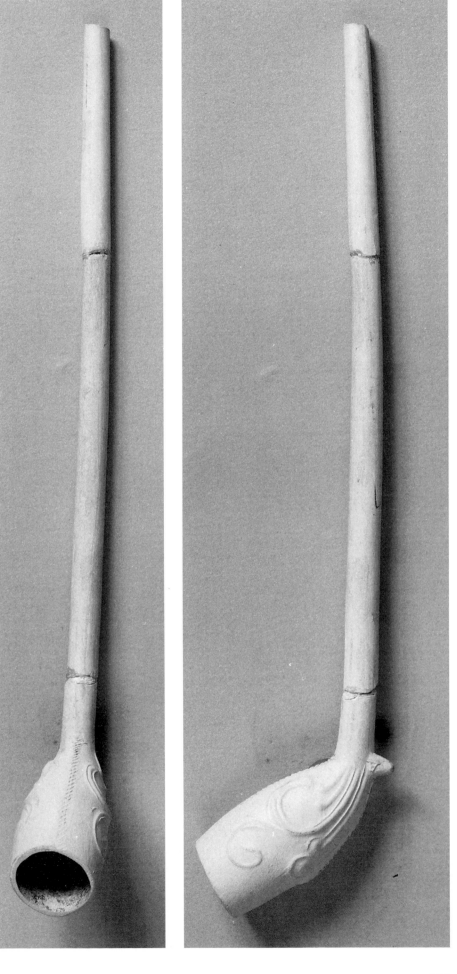

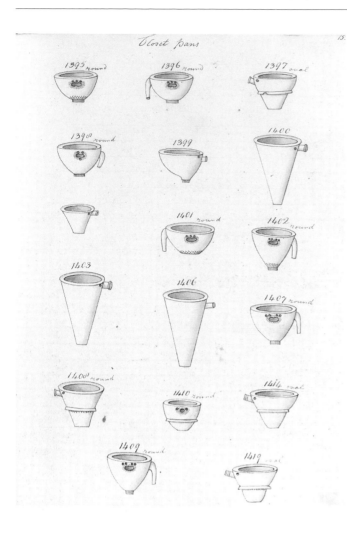

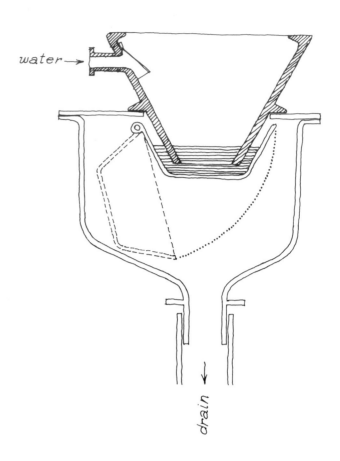

A page from an early 19th century Wedgwood shape book. The photographed example is probably no. 1397 oval. The Wedgwood Museum.

The pan closet in section with earthenware part shown hatched.

Subject: One of the earliest ceramic WCs. The photograph shows the ceramic part of a 'pan closet' which consisted largely of cast iron.
Manufacturer: Wedgwood, England.
Designer: Works design.
Materials: Earthenware, glazed over blue transfer.
Manufacturing processes: Hand pressing into plaster moulds. Drying and mould removal. Surface finishing. Drying. Biscuit firing. Decoration. Glazing. Final firing.
Dimensions: Long axis of the inner rim 42 cm.
Location: The Wedgwood Museum, Barlaston.
References & notes: *Wedgwood* vol. 1 & 2 by Robin Reilly, London 1989.
The Water Closet, by Roy Palmer, Newton Abbot.
Evaluation: Few historical design developments have been so curious as that concerned with human waste disposal. It started by being very simple and natural, as most of us know. Then with the development of the house and then the town, it became more and more complicated and less and less workable. Most mammals are intensely engaged in the disposal and hiding of their own waste. For the human being the problem has evidently been so pressing, that while he planned successfully in

other ways, he failed miserably in this respect. As soon as he moved his personal waste disposal away from the privy in the yard and into the house he made his first mistake. He tried to get rid of the waste 'automatically', as if by a wave of the hand, before he had acquired the necessary knowledge, foresight and technique for doing it. The problem in the towns was, of course, much greater than in the country but in both cases focus was put on the lavatory apparatus first and the planning of conveyance and final disposal second instead of the other way round; a procedure which typifies much of technological development.

The wash down closet shown here, though in itself an excellent product, played its part in this unfortunate historical mistake. It was a component part of a so-called pan closet which, although in use for a long period, was a complicated and fearfully unhygienic apparatus. It can be seen from the sketch that the cast iron funnel and pan simply became more and more soiled and infectious without any possibility of access for cleaning. Hygiene was greatly improved at the turn of the century by the cast-in-one WC with trap, and the septic tank, not to mention the great sewerage schemes.

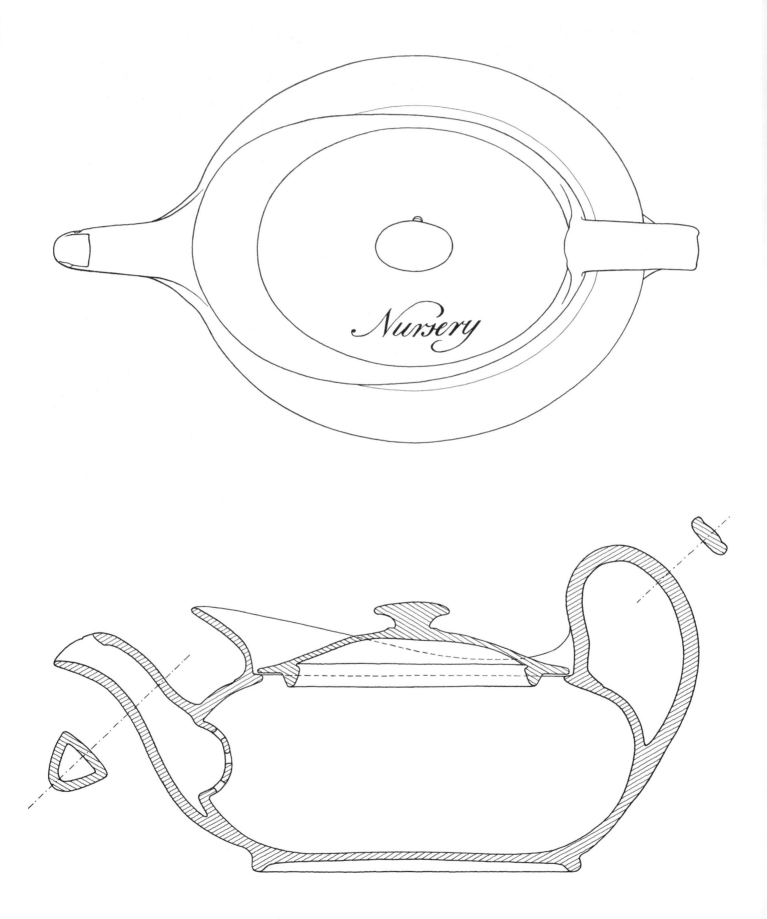

Measured drawing by Martin Bohøj.

CM 0 1 2 3 4 5 6 7 8 9 10

Photo S.R. Gnamm, Die Neue Sammlung, Munich.

Subject: The so-called 'parapet' teapot in a group of other 'useful wares'.
Manufacturer: Wedgwood, England.
Designer: Possibly William Wood who was apprenticed to and employed by Josiah Wedgwood.
Materials: Cream colour earthenware, or cream ware. Predecessor of white earthenware which is one of the most produced ceramic materials in England today. Cream colour earthenware was produced by English potteries as early as the beginning of the 18th century. Wedgwood's improved composition consisted primarily of 25% kaolin, 25% ball clay, 35% Cornish stone. Earthenware glaze with melting agent of lead silicate.
Manufacturing processes: Teapot cast in plaster mould. Handle, spout and knob sprigged to body and lid.
Dimensions: Length of teapot 22.8 cm.
Location: Kunstindustrimuseet (The Museum of Decorative Art), Copenhagen.
References & notes: *Britisk Brugskunst,* a selection of photographs from an exhibition of British products held in Copenhagen in 1933, described by the architect Steen Eiler Rasmussen.

It is not generally known that this exhibition and the book had a strong influence on Danish design.
Wedgwood vol.I & II, Robin Reilly, London 1989.
The Story of Wedgwood, Alison Kelly, London, 1975
Because of their importance as examples of early industrial design of very high quality the following two pages also deal with Wedgwood's useful wares.
Related products: Meat Carrier pp 82-83, Night Lamp and Warmer pp 84-85, Ewer pp 196-197.
Evaluation: One of the most fascinating aspects of industrial design history is the recurrent discovery of artefacts of the past which look as if they were designed yesterday. This occurs mostly in glass and ceramics – naturally enough for these are single-material products which have been fairly constant in their use and in their methods of making. The needs of fingers, hands, arms, and the consistency of food and drink have remained the same. Notice the serene relationship between the parapet of the teapot and the top of the jug – and their handles. The parapet acts as a funnel when adding the boiling water, and as a finger grip when pouring the tea.

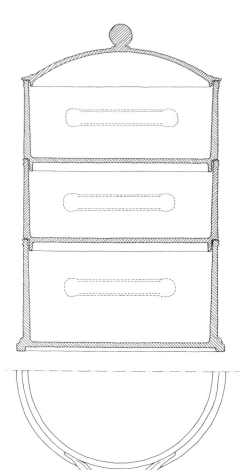

Drawn from a measured sketch. Scale 1:4

Subject: Stackable containers for taking hot food from the kitchen to the many rooms of large houses.
Design and manufacture: Josiah Wedgwood.
Materials: Cream colour earthenware.
Manufacturing processes: Cast in plaster moulds. Handles extruded, knobs cast. Handles and knobs sprigged to vessels and lid. These joints are placed by eye and modelled by hand.
Dimensions: Bottom container height 11.7 cm. Top and middle containers height 9.8 cm. Rim diameters 20.8 cm. Total height of stack 35.6 cm.
Location: The Wedgwood Museum, Barlaston.
References & notes: The meat carrier was part of a series of 'useful wares' which were produced throughout the first half of the 1800s.
Wedgwood vols. I & II by Robin Reilly, MacMillan, London 1989.

Related products: Night Lamp, next page.
Evaluation: This example of early Wedgwood embodies the essence of the functional tradition. Every industrial country has its examples of early industrial vernacular and they are often breathtaking in their simple sophistication. When they are at their best, as here, their design unites the user functions with expedience of manufacture to make a harmonious and practical whole. Here the stack, with its space saving and heat conserving advantages, consists of only three different elements: the base container with its expressive plinth, the second and third containers which are the same – and can be added to, with like containers – and the lid which can serve any of the containers. Rims are strengthened and junctions between elements defined by the fine double moulding. This also makes a shadow which detracts from any gaping there might be due to distortion under firing.

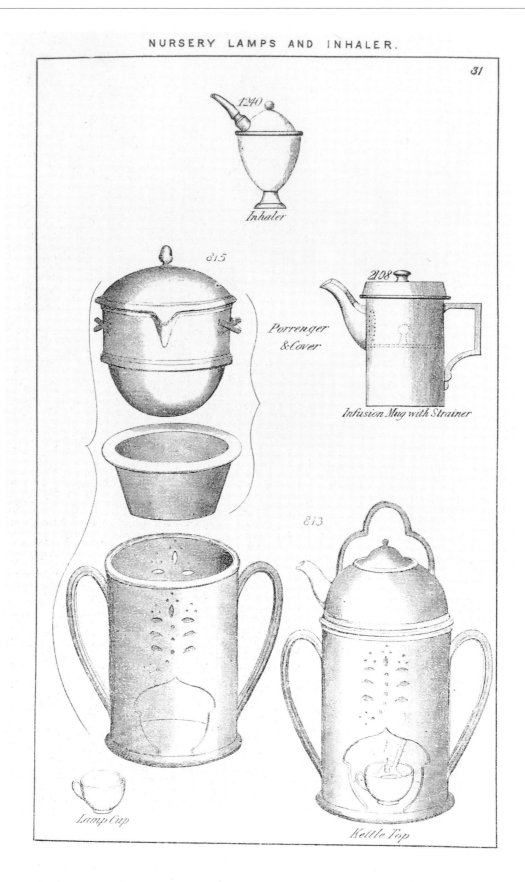

A page from Wedgwood's 1880 shape catalogue. The Wedgwood Museum.

This is another example of Wedgwood's 'cream colour useful ware'. It was for use in the bedroom or nursery for care of the sick at night. An oil lamp warmed the contents of a bowl, a teapot, a little kettle, or a porringer. The lamp also acted as a night light which illuminated the pierced plant decoration, and cast a flickering pattern of light onto the bedroom wall. This was a composite piece which provided several alternative services while performing a double function. This concept is an example of the beginnings of industrial product design, when advances in production techniques became synchronised with systematic design thinking. The height 27.5 cm. While the clay is leather-hard, the round perforations are cut-punched and the leaves and lamp opening cut with a knife.

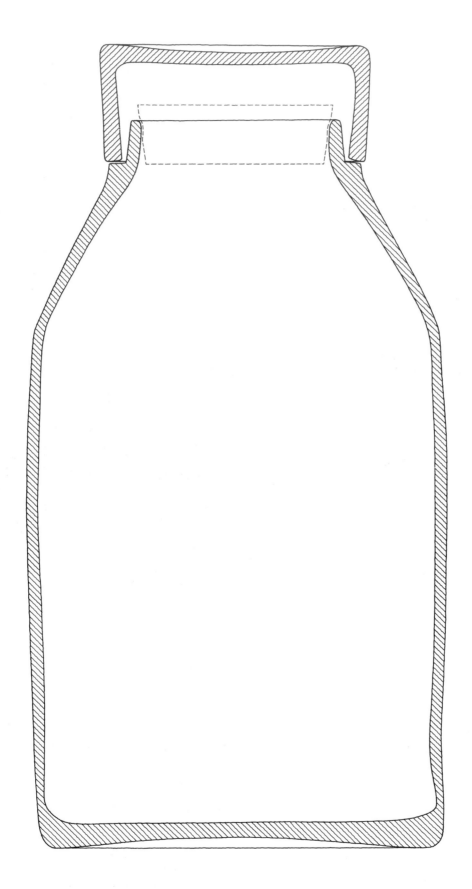

Measured drawing by Martin Bohøj. Scale 1:2

Photo: Gerry Johansson

Subject: These storage jars are very good examples of the Gamla (old) Höganäs tradition of Swedish pottery which lasted from the 1820s to the 1930s.
Designer: Probably the potter and his client.
Materials: Soft, straw-coloured earthenware. Clear lead glaze. Stoppers of cork.
Manufacturing processes: Hand thrown in series and fired to a temperature of about 1080°C.
Dimensions: The drawn example: height 42.5 cm. The smallest in the photo (and in the series) height 28 cm.
Location: Höganäs Museum.
References & notes: *Höganäskeramik, Konst-och bruks-föremål 1832-1926*, by Ann-Charlotte och Torsten Weimarck, ICA Förlaget, Sweden.
Related products: Chemist's Storage Jars, pp 192-193.

Evaluation: This is one of the few examples of lead-glazed pottery in the book. The technique was banned in Britain as early as the 1860s because of its toxic effect on both workers and consumers, but remained legal for a surprisingly long period on the Continent. The delightful and simple tradition of 'Gamla Höganäs' consisted of all sorts of household, kitchen, and utility pottery mostly in good, functional shapes. It was to be found all over Sweden and was regarded as standard equipment for many institutions and hospitals. These containers were for dry storage of powders, crystals etc. and made in five sizes. The largest was about a meter in height. The shoulders sloped up to a small neck which allowed the use of an air tight cork.

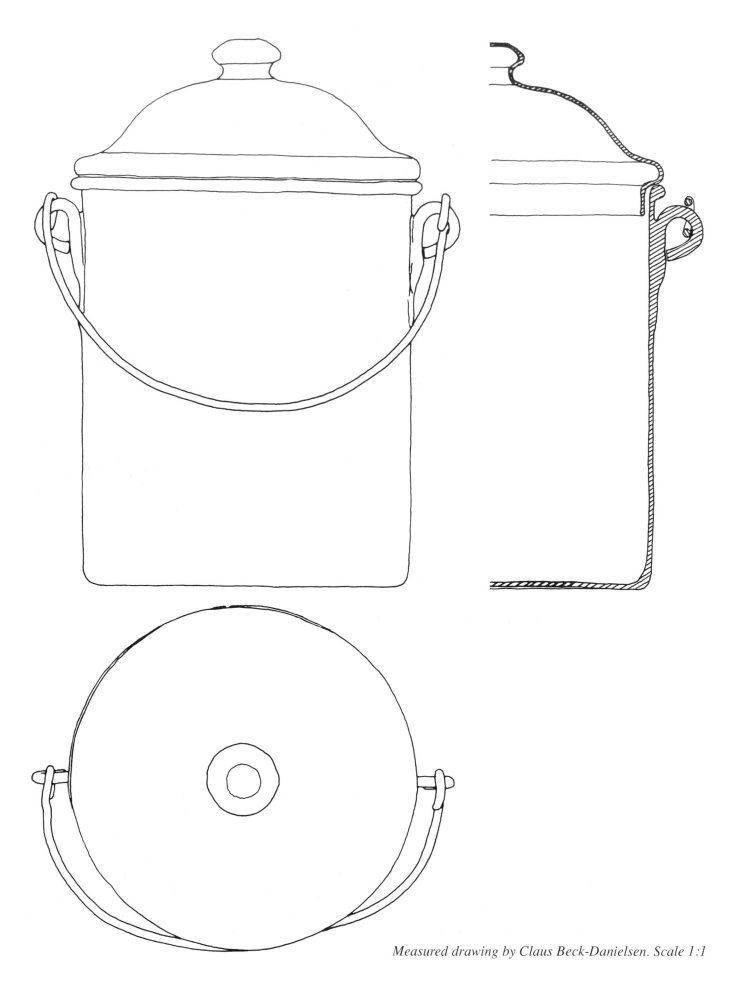

Measured drawing by Claus Beck-Danielsen. Scale 1:1

We have placed these items under Ceramic Products in order to stress the fact that enamel is a ceramic, or glaze, material. When it is applied to mild steel, which is its normal partner, the union of the two materials becomes, in effect, a third and quite different material. Enamelling as a protective surface was first introduced in the 18th century – inspired by the ancient jeweller's art. At the middle of the 19th century it started coming on to the market properly and has gradually developed to a very good material, though always brittle and chippable. Its resistance to heat makes it ideal for stoves and cooking utensils, it is resistant to dampness and acids, and it is easy to clean. Enamel can be applied in various colours and stencilled in the form of decoration or lettering. The chipped and rusting street signs and advertisements used to be one of the characteristic details of the street scene. The utensils shown here were pressed and/or spun; handle-eyes and lid handle welded or brazed on then dipped twice in boric-acid enamel slurry and fired after each dipping. The characteristic blue shade is obtained by adding cobalt oxide to the glaze, and the white interiors by adding tin oxide. See Re-use Cans, pp 102-103.

Measured sketch of B tile as in BOURNE STREET. Scale 1:1

Part of the display of various coloured tiles by Dunnill & Co. and Maw & Co. at the Jackfield factory, now part of the Ironbridge Gorge Museum.

From a terrace gable in Stoke-on-Trent.

Some of the finest details in many English towns are the street name signs done with tiles, let into the brickwork at street corners. The wall tile, both indoors and out, both decorative and plain, was part of the 19th and early 20th century scene and these street name tiles were an extension of this. They were made in the same way – dust pressed – and the letters, one per tile, either inlaid in a different clay colour or transferred beneath the final glaze. The use of tiles for street naming is an excellent practise. The sharp contrast between the white letter and the dark background is both noticeable and legible. The signs are hard to damage, weatherproof – in fact cleaned by the weather – and there is a quality of permanence about them. They can be ruined by a bad letterface but fortunately among the many manufacturers who made tile alphabets there were good letterforms. 'BOURNE STREET' is a particularly nice one.

The dust pressing technique of the 1840s for the manufacture of tiles was one of the ceramic industry's major inventions. It was the solution to the production of dimensionally identical, flat tiles. Clay in its workable, damp condition is an unstable material. It shrinks and distorts during drying, as do most materials that contain water, and continues to do so during firing. The finished tile has to fit edge to edge and form an even surface over existing walls with their corners, reveals and openings, and the tolerances are small. The dust pressing process, as its name implies, shapes the clay, while it is in the form of almost dry dust, under high pressure between cast iron moulds. This is followed by the firing and glazing processes during which shrinkage and distortion are so slight that they can be calculated and controlled. Dust pressing was the technique which helped the tiled surface to fulfil its function as a clean and washable surface.

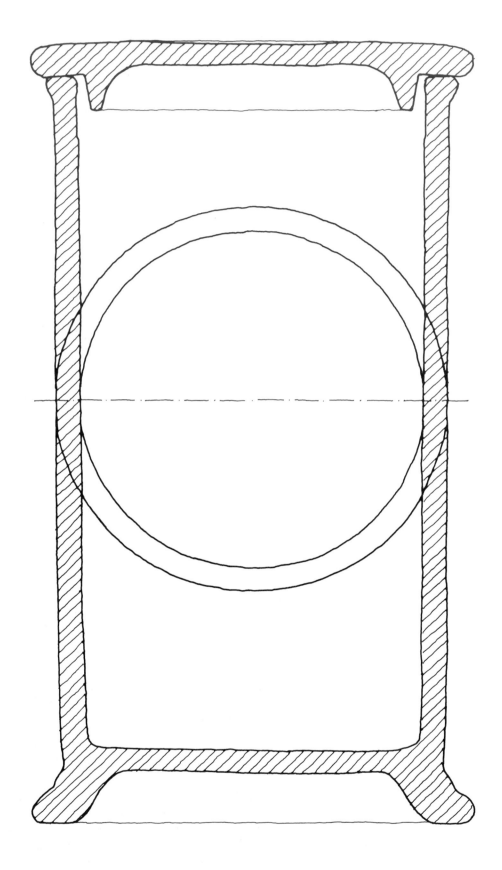

Measured drawing by Nis Øllgaard. Scale 1:1

Subject: Dry storage jars for chemists' shops.
Manufacture & design: Unknown to the authors, probably German.
Materials: Clear glazed technical porcelain with lettering in black over-glaze.
Manufacturing process: The jars are slipcast in multiple-part moulds with removable bases for making the raised bottoms. Lids slipcast in double moulds. Clear glazed except for standing surface and where lids and jar rims meet. Hand lettering.
Dimensions: Capacity 1 litre. Weight 1275 g.
Location: Author's collection.
References & notes: The Open Air Museum, Den Gamle By (The Old Town), Århus, has collections of similar jars which are hand decorated.

Related products: Chemist's Containers pp 186-187.
Evaluation: Jars such as these have been familiar in chemists' shops in northern Europe throughout the 19th and for much of the 20th centuries. With their clearly expressed function, and when displayed in rows on the chemists' shelves, they were one of the characteristic elements of the shop interior. They are a very good example of a strong, inexpensive, quantity produced, basic form which could be supplied with, or without, a choice of decoration. Here the hand-painted labels provide a functional decoration, as do the jars themselves. Lids meet the jars with unglazed, non-slippery surfaces. Because of the girth and weight of the jars they have to be lifted with both hands. For comfortable holding with one hand, they would have to be at least 1.5 cm less in diameter.

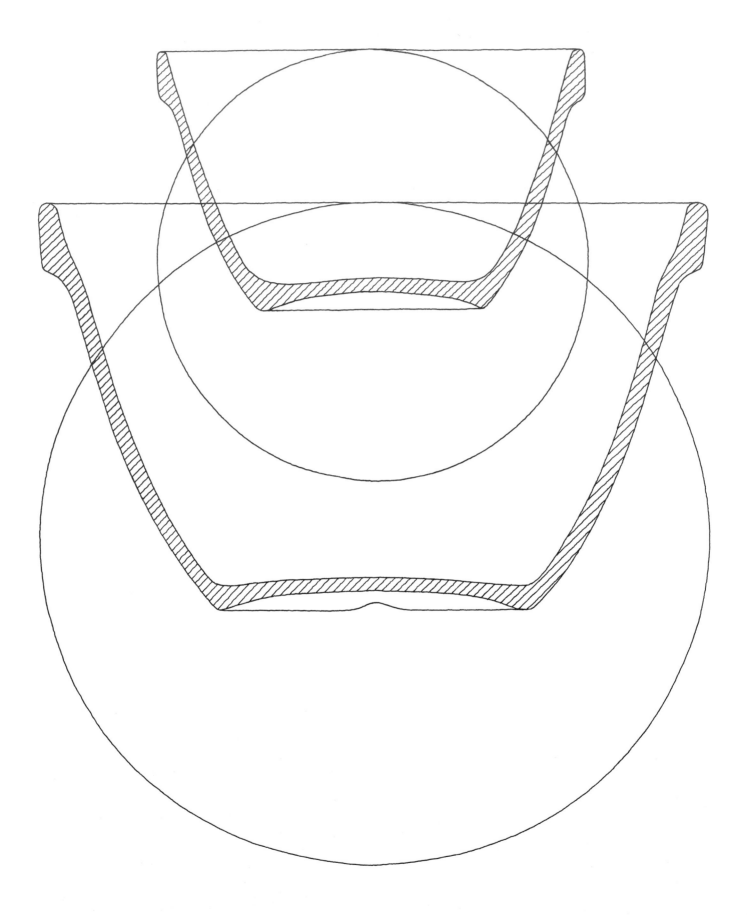

Sections and plans showing ½ pt and 2 pt sizes. Measured drawing by Martin Bohøj. Scale 1:1

Subject: Basins in which to steam puddings.
Manufacturer: T.G. Green Limited, England.
Design: Vernacular.
Materials: Clear glazed earthenware.
Manufacturing processes: Interior and rim turned with template while the clay is leatherhard and still in its outer mould. This much used process is called 'jiggering'.
Dimensions & capacities: 4.5 in. diameter 0.5 pt, 5.0 in. diameter 0.75 pt, 5.5 in. diameter 1.0 pt, 6 in. diameter 1.5 pt, 7 in. diameter 2.0 pt, 7.75 in. diameter 2.5 pt.
Location: Author's collection.
References & notes: *Good Housekeeping Cookery Book*, Gramol Publ., London 1948.
Related products: Kitchen and Dining Set pp 202-203
Evaluation: The English pudding basin is a good example of a traditional utensil of daily use whose design is wholly determined by its exclusively national function:

the making of steamed puddings. The steamed pudding is normally a sweet dish composed of flour, bread crumbs, baking powder, salt, suet, sugar, milk and the flavouring ingredient from which the pudding often gets its name, e.g. chocolate pudding, marmalade pudding etc. There are over 30 different types – not forgetting Christmas pudding. The mixture is emptied into a greased basin, a cloth stretched over the top and tied with string under the projecting rim and steamed. The pudding basin gets its characteristic shape from the necessity for the pudding to slip free when turned upside down onto the serving dish. The basin is thus the mould which gives the Christmas pudding, and all the other puddings, their well-known form. Pudding basins have been made by many English potteries at very low cost – a Cinderella of ceramics unheeded by its inventors and users but, because it belongs to a national vernacular, treasured by foreigners.

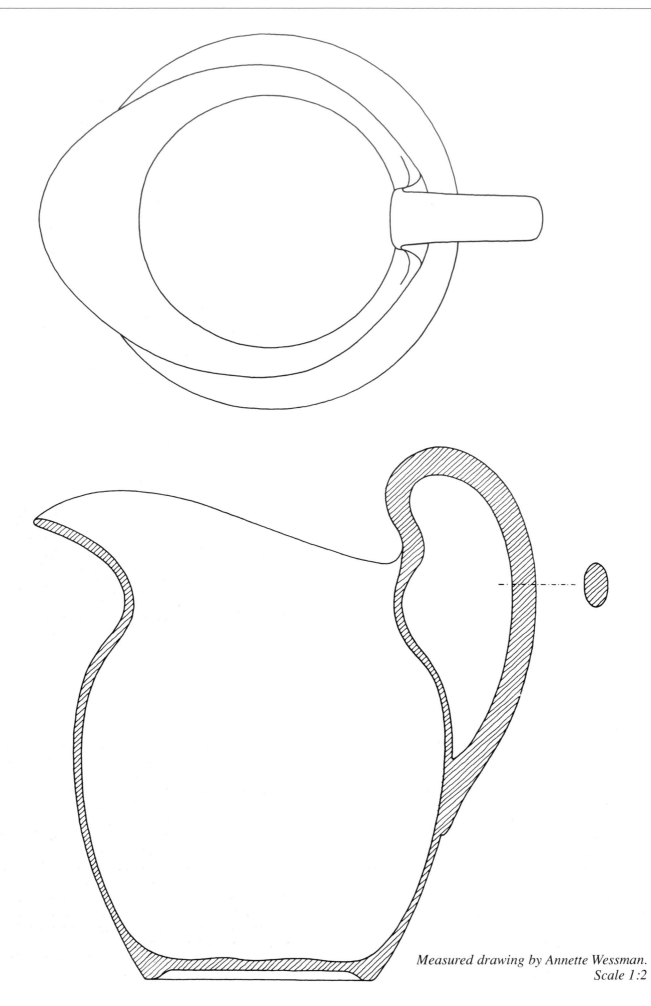

Measured drawing by Annette Wessman.
Scale 1:2

Subject: This water ewer for personal washing will have belonged to a wash-stand set including a wash basin, in which it stood while not in use, and a soap tray.
Manufacturer/designer: Unknown to the authors but typical of a large production of inexpensive ware from Stoke-on-Trent, England.
Materials: Earthenware with clear glaze.
Manufacturing processes: Slipcast in multiple part mould. Handle cast and sprigged to the body and hand modelled where it joins the rim. Rim, hand-finished in the mould while clay is still soft.
Dimensions: Height 28.5 cm, capacity 4 litres.
Location: Author's collection.
References & notes: Will have been supplied plain, as here, for institutions etc., or decorated.
Related products: See similar shape, shown with Wedgwood Teapot pp 180-181.

Evaluation: This is one of the best examples of integration of form and function known to the authors. The design is so exceptional that it cannot merely be classified as vernacular. A good potter must be responsible for the form, and Wedgwood was making smaller jugs like this in the 18th century.

After using the jug over a long period one is convinced no other container, spout and handle could be designed that would unite more pleasingly or more aptly the jobs they have to perform: the container is shaped to hold a large quantity of water as compactly as possible, the neck goes elegantly but firmly over into a 'funnel' and spout which first receive and then deliver water in an even, non-splashing stream into the wash basin. The handle grows out of the rim in such a way as to enable the holder to adjust to the changes of weight whilst either filling, carrying or pouring.

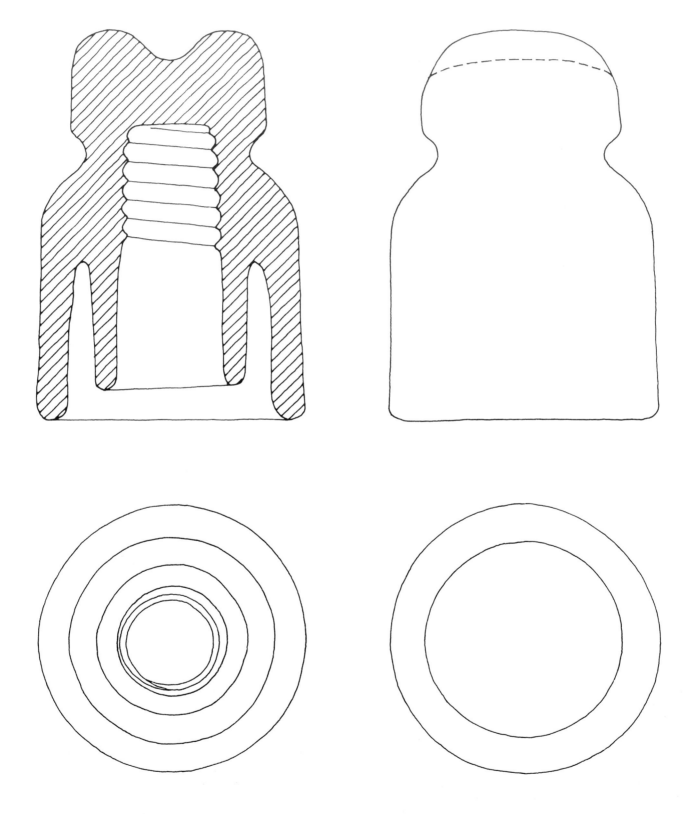

Measured drawing by Claus Beck-Danielsen. Scale 1:1

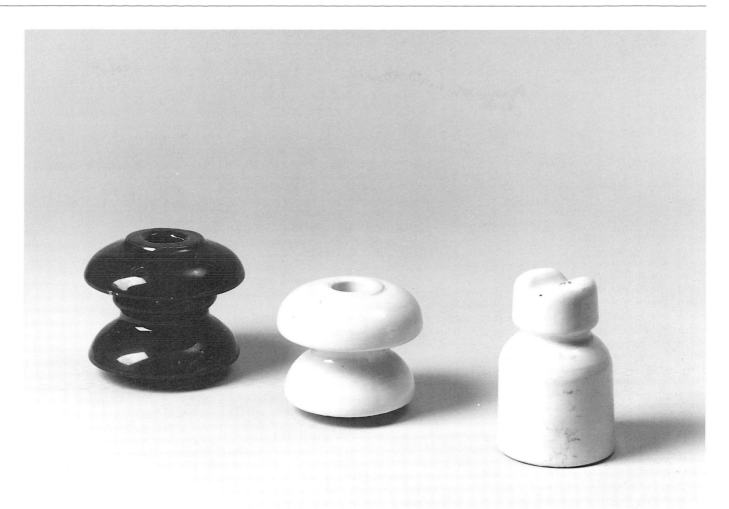

Subject: Mast insulators for high voltage and (on far right and drawing) a domestic insulator.
Manufacturer: Porcelænsfabrikken Norden, Copenhagen.
Designer: The electricity authorities in conjunction with the manufacturer.
Materials: Industrial porcelain.
Manufacturing processes: Turned in solid piece of leather-hard porcelain. Internal screw thread tapped. Glazed at first firing.
Location: Author's collection.
References & notes: The Danish Museum of Electricity and Gudenå Centre, at Tange.
Evaluation: These insulators, and the lamp on the following page, are perhaps the best examples there can be for showing the fundamental role which the ceramics industry played as a service industry in the development of technology. It is hard to imagine electrification without the use of porcelain to confine and control current. The whole electrical process from generation right through to consumption has been, and still is to a lesser extent, served by thousands of different shapes and sizes of porcelain components. The three insulators shown here,

under 1900-1910, are in fact from the last half of the 20th century but they are in principle similar to those made at the beginning of the century by the same manufacturer. The basic requirements have not changed much in nearly 100 years. Porcelain was chosen from the beginning for the distribution of electricity. It has high electrical resistance which entails resistance to both water absorption and the accumulation of surface water and dirt, and it is sufficiently robust to withstand the considerable loads imposed on it in all temperatures and weather conditions over very long periods. The design of the insulator, as can be seen from the drawing, has to enable the material to fulfil these requirements. It is only at the very end of the 20th century that the mast insulators have begun to lose their importance due to telegraph and electricity cables being laid underground.

A techno-cultural curiosity is that porcelain, discovered 2000 years ago in China and re-discovered 1500 years later in Germany as a semi-precious, decorative material, should also become a utility material indispensable to electrification and to industrialisation in general.

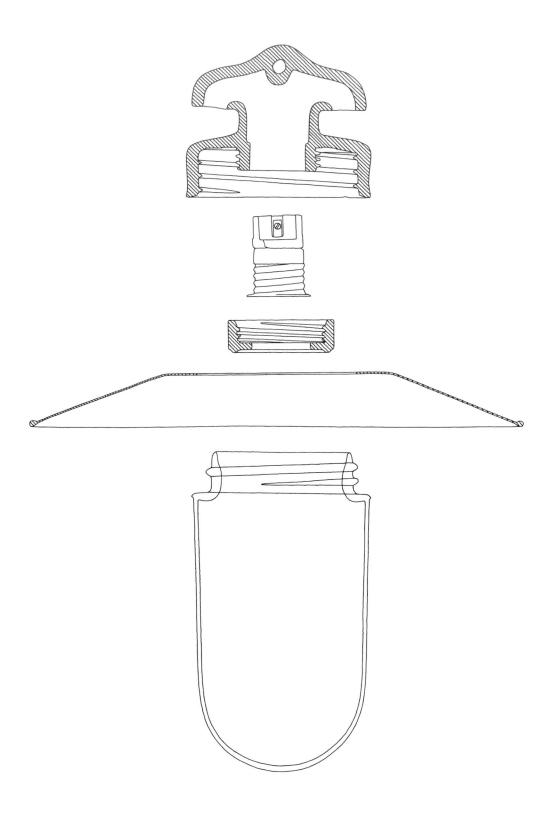

Measured drawing by Martin Bohøj. Scale 1:2

Subject: Hanging lamp for stables, pigsties, etc.
Manufacturer: In the catalogues of Aksel Skov a/s, Denmark.
Materials: Industrial porcelain, glass globe, enamelled sheet steel shade, bulb socket of copper and porcelain.
Manufacturing processes: Porcelain for main housing, bulb socket fixing ring, and terminal: cast in moulds. Screw threads tapped in the porcelain while in its leather-hard state. Globe, and its screw thread, pressed glass.
Location: Museum of Electricity, Denmark.
Related products: Insulators, on previous page.
Evaluation: The Danes were quick to see the potential importance of electricity to agriculture. Signs of this can still be seen in the form of small, and now rather quaint, tower-formed transformer buildings here and there in the landscape, and many remember this pendant lamp hanging in rows in the pigsty. It is a good piece of utility design consisting of the minimum number of components all held together by screw threads in porcelain and in the glass globe. Parts could be replaced. Concern as to the safety of electricity is shown in the careful separation of the positive and negative wires of the flex. The damp conditions of farm buildings are well guarded against with the intricately formed parts of porcelain.

Subject: Six examples of various generations of 'Cornish Kitchen Ware'. The series comprises a complete set of utensils for cooking and dining.
Manufacturer: T.G. Green Limited, England.
Designer: Works design, later re-designed by Judith Onions. The drawn storage jar is by her.
Materials: Earthenware. Lid has rubber seal.
Manufacturing processes: The blue slip-covered pots are turned when the clay has become leather-hard to expose bands of the white body.
Location: Author's collection.
References: The manufacturer.
Related products: Pudding Basins pp 194-195.
Evaluation: Cornish Ware started in 1920 with utensils for the kitchen such as storage jars; later it was increased to some 40 pieces for kitchen and dining. It is still in production over 70 years later with 30 pieces. Its appearance on the market coincided with the disappearance of the paid cook in middle-class homes and the beginnings of the labour-saving kitchen and its fusion with the dining space. In this way Cornish Kitchen Ware became part of a new house planning concept and its instantly recognisable, optimistic character reinforced this. The shapes and their decoration are largely dictated by the processes of turning on the lathe which gives the series certain design qualities of simplicity and practicality. Where the turning tool has shaved the slip away, there is a little step between white and blue bands giving precision and relief to the surface.

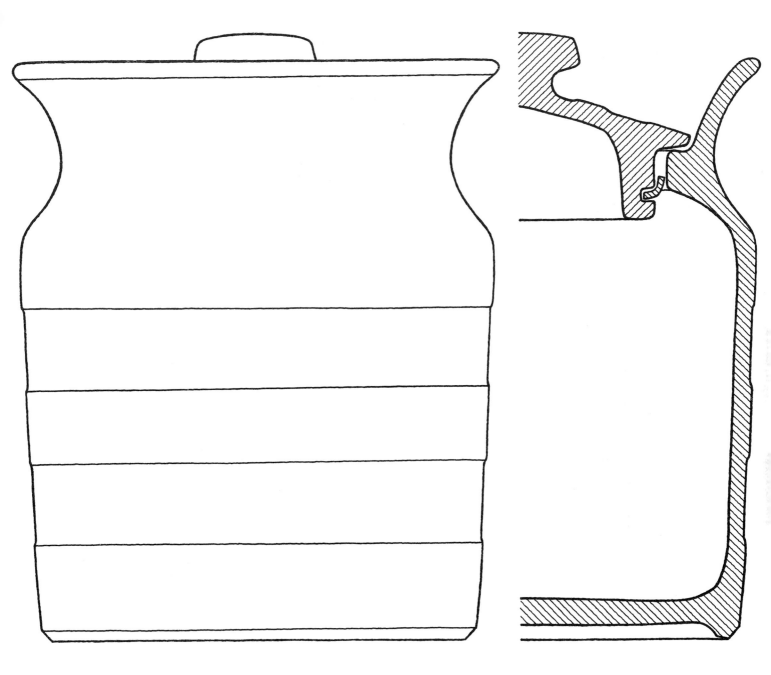

Measured drawing by Martin Bohøj. Scale 1:1

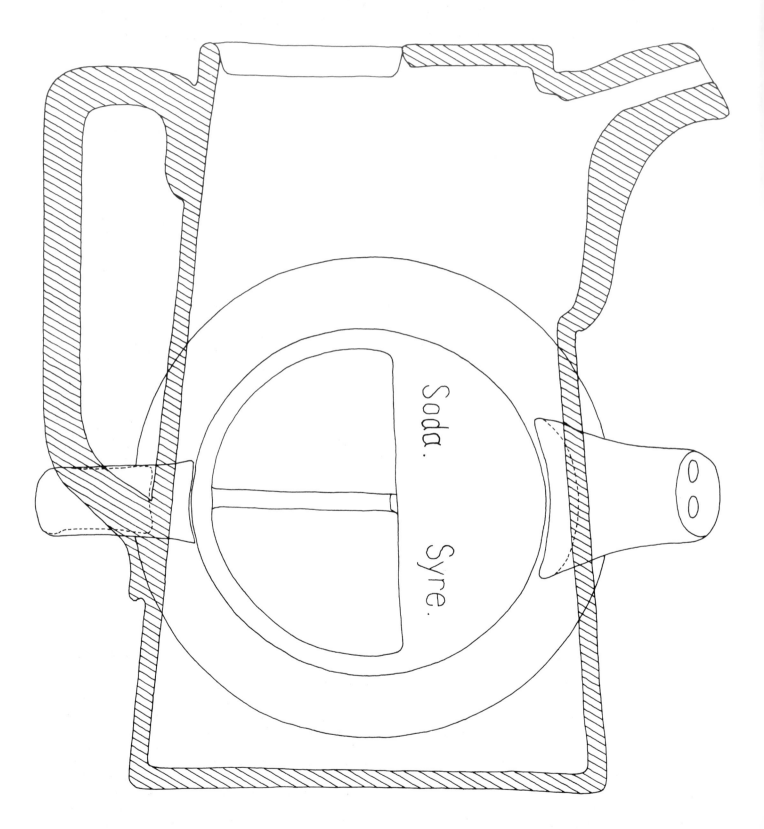

Measured drawing by Claus Bech-Danielsen.

Photo: Louis Schnakenburg

Subject: Mixing jug for making fizzy lemonade. Also shown are measuring jugs and photographic developing pans.

Manufacturer: Aluminia, Denmark.

Materials: Clear glazed earthenware.

Manufacturing processes: The jug is cast in a conical form without the bottom. The internal partition is placed and sprigged, followed by the bottom. The holes in the spout are pierced out with a metal tube. The handle is cast separately and then sprigged in place.

Dimensions: Actual height 22 cm.

Location: Private collection.

Evaluation: What could almost be described as a Danish national drink called *sodavand*, which consists of soda water and sugar with various fruit flavours, has been made and bottled by all the main breweries for many years. Like most manufactured foodstuffs *sodavand* originated as a homemade drink and this jug was designed for making it in. Citric acid, sugar and water were poured into the compartment marked 'Syre'; sodium bicarbonate, sugar and water into the compartment marked 'Soda'. On pouring out the concoctions, equal parts became mixed in the glass and ready to drink. Later on, the powders could be purchased ready-mixed in pairs of packets, one to be poured into each compartment and water added. This simply conceived and practical utensil is part of a historical process which was typical for the industrialisation of foodstuffs: first user procurement of ingredients and their preparation; then factory production of ingredients and user preparation; then factory production and distribution of the single product. In truth it is hard to understand how home making, with the help of this single piece of simple equipment, could ever have been replaced by the whole complex of industrial brewing with its machinery for dispensing, mixing, pasteurising, bottle making, bottling, capsule making, sealing, label printing, labelling, crate manufacture, crating, transport, empty bottle collection and washing ...!

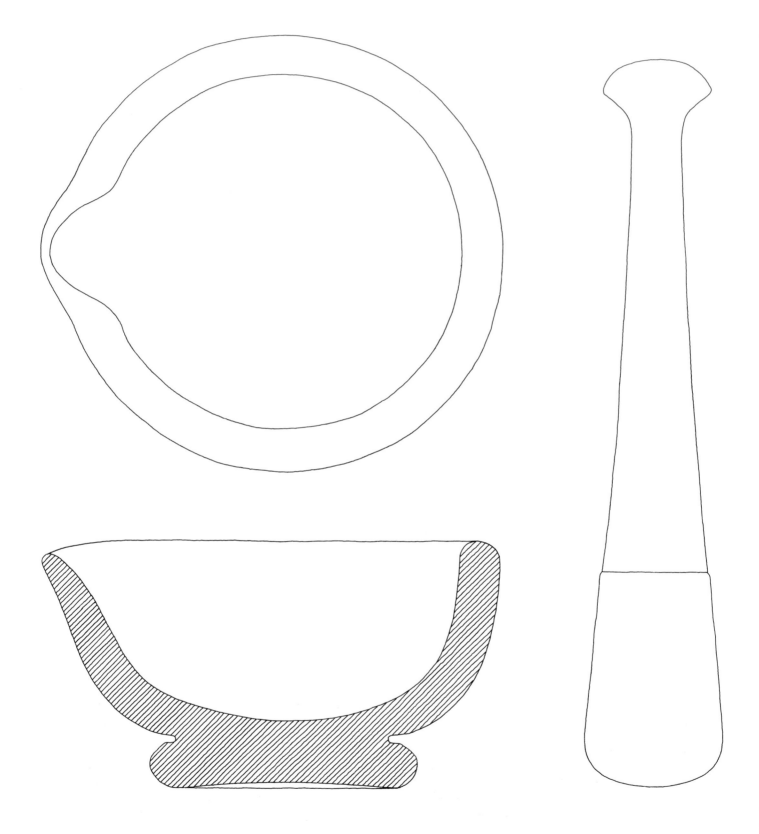

Measured drawing by Nis Øllgaard. Scale 1:1

Subject: Mortar and pestle.
Manufacturer: Milton Brook Pottery, Stoke-on-Trent, England. First produced 1933.
Designer: Works design.
Materials: Unglazed porcelain and beech wood.
Manufacturing processes: Mortar: thrown with profile-turned foot. Pestle: head turned by potter, handle turned by wood turner. Probably joined with round mortise and tennon (not shown).
Dimensions: Size no.1, 4.5 in. (11.5 cm) external diameter. Made in six sizes from 3.5 in. to 14 in.
Location: Author's collection.
Evaluation: The mortar and pestle is one of man's oldest tools – the basic mill. The pulverisation of substances, both edible and non-edible, has been and still is the function which has demanded man's greatest mechanical ingenuity. And yet the original method, with its simplicity and manifold uses, remains universal, although the material used – clay, glass, metal, wood or stone – will depend both on the desired result and on what material is at hand. This example is a very convincing utensil: it stands firmly and enables powerful grinding. The unglazed surfaces of both mortar and pestle head give a good grinding friction. The pestle's unlacquered wooden handle with its mushroom formed end allows good grip, control and pressure. The mortar and pestle is also a potter's tool so the excellence of this one is perhaps not surprising.

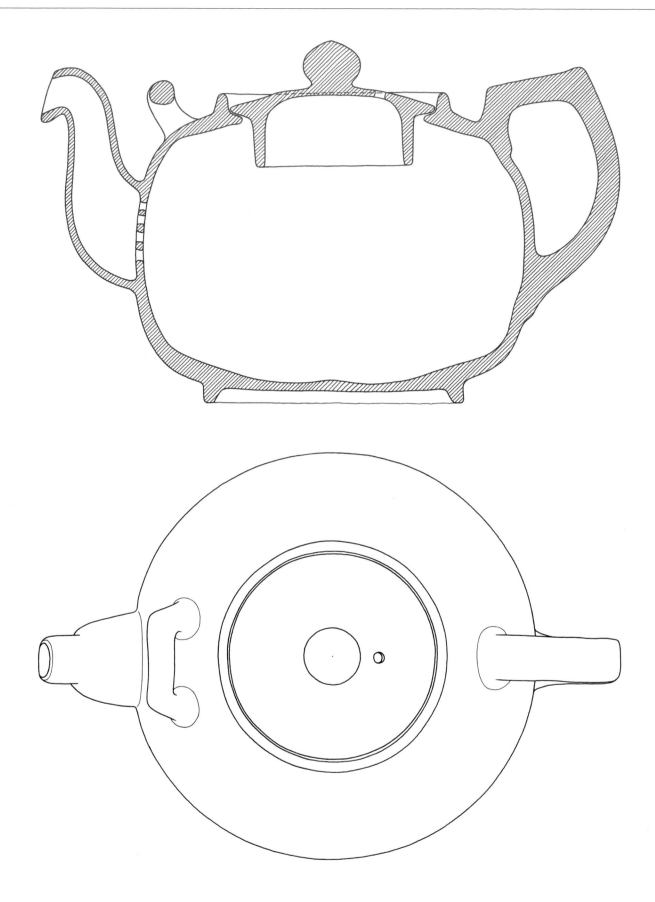

Measured drawing by Martin Bohøj. Scale 1.2

Subject: Large teapot for the catering trade.
Manufacturer: The Denby Pottery Co., England.
Designers: Shape, W. Mitchell and J.Ransden;
Glaze, D. Gilbert.
Materials: Stoneware clay. Glaze outside, brown iron
oxide. Inside, pale blue cobalt oxide glaze.
Manufacturing processes: Thrown on a potter's wheel.
Handles, knob and spout cast and sprigged.
Dimensions: Length 31 cm. Capacity 6 pt (3.4 *l.*)
Made in seven sizes from ¾ pt to 8 pt.
Location: Author's collection.
References & notes: Information supplied by the manu-
facturer. The shape, which is named 'Nevva Drip', is

from 1922 and the characteristic brown glaze was introduced in the 1940s. There are in all ten sieve perforations
arranged symetrically.
Related products: Teapots, pp 180-181. and pp 260-261.
Evaluation: This teapot is in every way a remarkably
robust piece of pottery. Its weight alone – over 2.5 kg –
gives an idea of its physique. With its specially
developed glaze and its strength of form, it has become
well-known and almost synonymous with good, strong,
English tea. The non-drip spout detail almost lives up to
its name, and is a good solution to one of the potter's
greatest problems. It also maintains strength just where
the teapot is most exposed to damage.

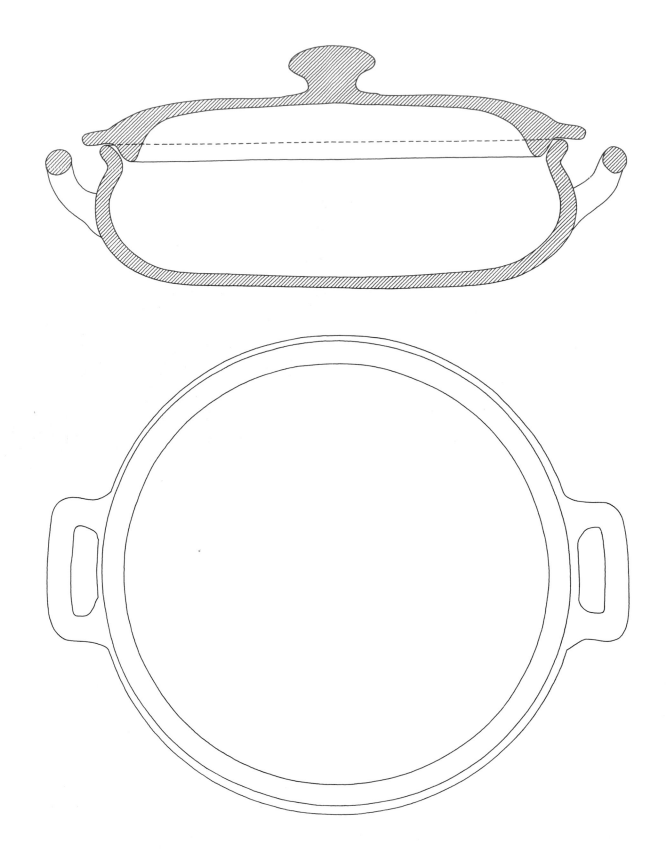

Measured drawing by Martin Bohøj. Scale 1:2

Subject: Oven and flameproof casserole from a series of fireproof ware.
Manufacturer: Vulcania, Siena, Italy.
Design: Traditional.
Materials: Red earthenware of siliceous, aluminous and ferrous clays. Clear frit glaze.
Manufacturing processes: Cast in plaster moulds, handles sprigged, first firing probably between 800°C and 900°C, glost 1050°C to 1080°C.
Dimensions: Outer diameter of pot rim 24.5 cm.
Location: Author's collection.
References & notes: Information supplied by Vulcania (founded 1911). This type of pottery is typical of a long tradition of useful ceramics from Italy. The underside is unglazed.
Related products: 'Ildpot', see next page.
Comment: A word about the properties achieved by different clay mixes and firing temperatures. The fireproofness of this series and its resistance to temperature change is due to the slightly porous clay which is able to absorb the tensions which occur in the body during sudden heating and cooling. This is a property inherited from our forefathers whose cooking utensils consisted solely of simple earthenware fired at low temperatures. Compared with the fireproof series 'Ildpot', discussed on the next page, it can be seen as being worlds apart. The casserole is a relaxed, soft pot which merely makes a thud when tapped – no ring. 'Ildpot', with its much later technology, rings like a bell and is hard in body and precise in form. It is a high energy and rescource-using product to make, compared with this casserole, and yet they each perform many of the same functions. We have placed the casserole in the 1960s not only because it was being produced at this time but also to illustrate the timelessness of pottery. In principle, it could not only have been placed anywhere in the book, but anywhere in the last 2000 years. The art of ceramics is one of the oldest and most constant we have.

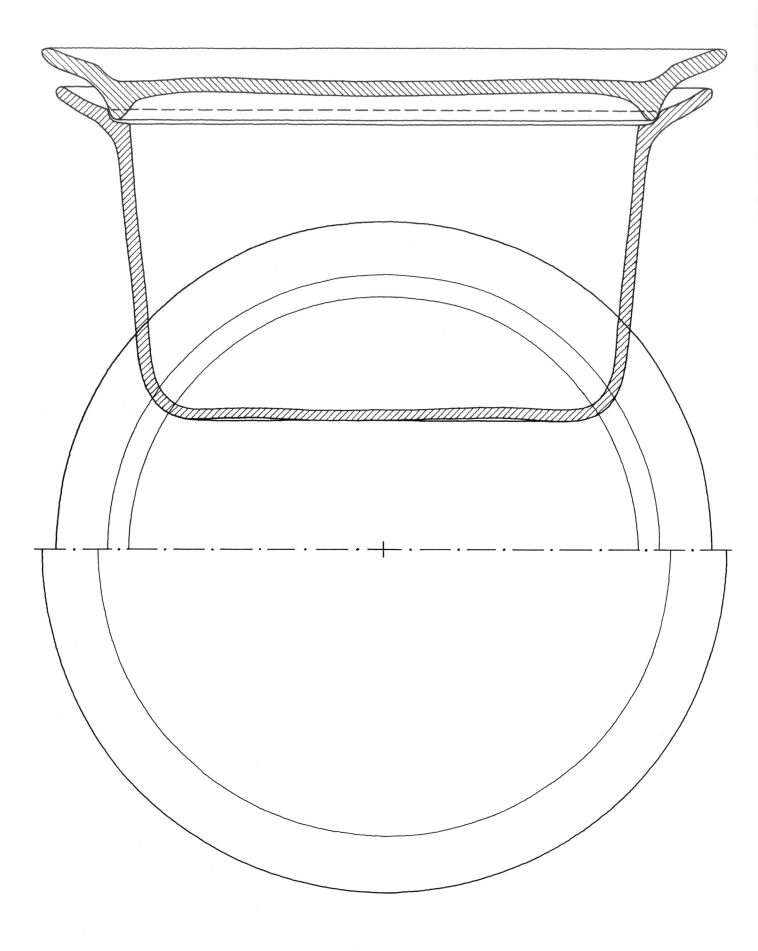

Measured drawing by Martin Bohøj.

Subject: A series of 17 flame and oven proof pots.
Manufacturer: Royal Copenhagen Porcelain.
Designer: Grete Meyer.
Materials: Stoneware of cordierite which is a plastic mineral consisting of a silicate of magnesium, aluminium and iron.
Manufacturing processes: Lid: turned on a plaster of Paris mould like a plate, base formed by template. Bowl: turned in plaster mould, inside formed by template, a process known as 'jiggering'. Fired at 1430°C.
Dimensions: Photograph from left to right, lids excluded: 20.5 x 29.5 cm, height 6 cm;
Diam. 25.5 cm, height 14.5 cm; diam. 13.5 x 5.5 cm; diam. 25.5 x 10 cm; diam. 21.5 x 11 cm. (drawing).
Location: Author's collection.
Related products: Casserole, see p. 211
Evaluation: 'Ildpot' is a series of oven and flameproof casseroles, basins and dishes that are able to withstand great contrasts of temperature – from the refrigerator direct to the oven. The cordierite stoneware from which it is made is remarkably light but intensely hard and strong. It is a vitrified, unglazed product which has an attractive warm grey, matt surface. When handled, it gives off a resonant ring. A slight colour variation between pots increases with use and becomes even warmer in tone. The design of 'Ildpot' seems to be in complete harmony with the character of this unusual material. In planning the series, the types of cooking for which it should be used, the dimensions of domestic ovens and the sizes of gas rings, electric plates and spirals have all been taken into consideration. Distortion in the furnace prevents the attainment of true plain surfaces so Ildpot is best suited to gas rings. The shapes are exceptional: they do not have the usual two handles but a generous rim which acts as a handle all round; in the oven, on the ring and on the dining table all pots can be lifted with two hands (and an oven cloth) from any angle. Vessels are well rounded with constant wall thicknesses which makes them good to work with and reduces tensions being set up under extreme temperatures. Lids have many uses, as plates, as serving dishes, as stands and for grilling on. 'Ildpot' is unique in that it is technically advanced and precise in form and at the same time embodies the living qualities of the good traditional pot.

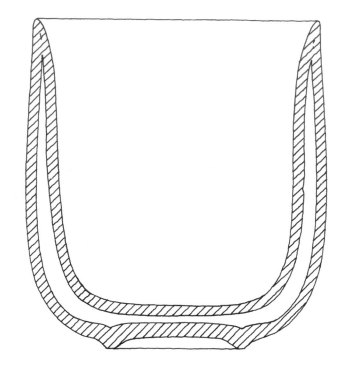

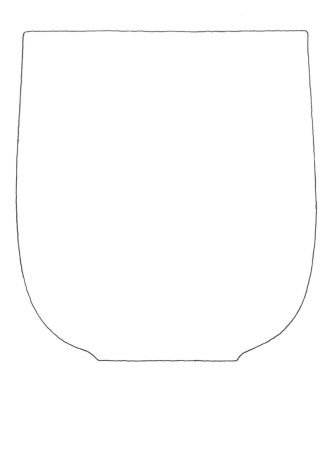

Subject: Thermos beaker.
Manufacturer: Snorre Stephensen. These beakers have been in production since 1970.
Designer: Snorre Læssøe Stephensen.
Materials: Porcelain with transparent feldspar glaze. Firing 1260°C. Oxidising.
Manufacturing processes: 'Termokop' consists of two thin porcelain shells which are each cast separately in plaster moulds giving a body thickness of about 2.8 mm. While in leather-hard condition, the shells are placed one inside the other and sprigged at the rim. The 'double' wall at the rim is turned to about 1 mm thick and formed to a comfortable drinking edge. To achieve optimal thermos effect it is important that the two porcelain shells touch at the top edge only. Handling of the thin shells in their various conditions, and joining and turning them, demands skill and judgement. 'Termokop' is produced in a small workshop which specialises in utensils of cast porcelain.
Evaluation: Here the double function of keeping the drink hot and the beaker cool to hold, is served, without any form of handle. This works remarkably well. The beakers have a very special quality which is, at once, visual and tactile: their robust form contrasts with their fine rims and their surprising lightness in weight.

Meaured drawing by Martin Bohøj. Scale 1:1

The inner and outer shells shown in the half parts of their plaster moulds. The mould for the outer shell, right, divides horizontally so that it can be separated to allow for the 'returning' shell form.

PART V

GLASS PRODUCTS

The word 'glass' comes from the Latin 'glesum', which means 'amber'. The material consists of sand, soda, chalk, potash and sometimes lead, which when melted at 1200 – 1300°C becomes a syrup-like mass of silicates. It has a very high viscosity which enables it to be made into large bubbles, or drawn out into long, thin threads, or flow out and become thin sheets when poured onto casting trays. After cooling, glass maintains its amorphous character and, as though it were solidified water, lets light pass through unhindered. This is unique in relation to other melted materials which always re-crystallise, and therefore become more or less opaque. Chemists often describe glass as 'supercooled liquid', and research continues into why the material possesses this special property. Glass is known principally for its transparence, but besides this it is very hard, dimensionally stable, and resistant to chemicals.

We do not know when man learned to make glass. The oldest finding so far, in Persepolis in Asia Minor, dates glass to 4000BC. Here glass is found as glaze on ceramics, in the form of brightly coloured bricks in turquoise, yellow and green. The oldest quantity produced glass containers were found in the Egyptian tombs and are dated between 1500 and 1000BC, and we know how they were made. A core of sand and organic binder was formed round a metal rod which was then dipped into molten glass. When the glass had hardened around the core, the sand was knocked out leaving a hollow glass container. This 'sandcore' technique existed for hundreds of years. Then the glass-blowing technique was invented, and the metal rod was replaced by a pipe, on which the molten glass was collected in a clump on one end. The hollowness in the glass comes when the glassmaker blows through the opposite end. As far as is known, the technique was adopted in about 100BC and came to be the foundation of the craft of glassmaking, and its development in Europe, for the following centuries. The dominant position which the Romans held in glass production was based on the techniques which the glassmakers of Asia Minor and Egypt had already developed.

Production in the glass industry follows three main principles: *blowing, pressing, and casting.*

In *mould blowing* a glass bubble is blown into a metal or wooden mould. The work is either done manually or as a fully automatic process. Some sculptural and decorative glass, bottles and drinking glasses are made in this way.

GLASS PRODUCTS

Free blowing (meaning blown without the use of a mould) is used for large glass products which are either too complicated or impossible to make with moulds. Series of products whose design and quality are dependent on glassmaker techniques, without the use of moulds, are also free blown. Free blowing allows greater freedom of shape and inspires the glassmaker's virtuosity. It requires good judgement of size and proportion, and a feeling for timing, since the hot glass only has a short period of workability, before it has to be re-heated.

Pressing is the commonest way to produce glass articles up to a certain size. Pressing is done in metal moulds, with a plunger which forces the glass into the mould. The process gives a high degree of precision and dimensional accuracy, and offers the possibility of surface pattern. Pressing is often used for making both domestic and technical glass.

Glass articles can also be made by the use of *centrifugal force*. The molten glass is flung out into a mould of the required shape. The mould leaves its impression on the outside of the glass, whereas the inside retains its untouched, glassy character. The technique is used in the production of certain special glasses and wine glasses.

The *casting* of window glass has been known since ancient times as a simple, primitive way of making flat glass. At the end of the 17th century the technique was improved by the introduction of a roller to give the glass an even thickness. The hot glass is poured onto a plane steel bed with upstanding edges and rolled. This, in principle, is how plate glass is made. The casting process is still used to make the thin window glass. Many solid articles are cast. This is done in the traditional way in metal or sand moulds. The method is similar to metal casting, but with the difference that thick glass requires a long cooling time to lose its thermal tensions.

Today the production of flat glass has become a highly developed industry in which casting is done in a continual process. The glass flows from the furnace, through a matrix, which controls its thickness, and onto long steel cooling beds. 'Float' glass gets its name from the cooling process whereby the hot glass is floated onto molten tin. This gives the glass a flat, smooth surface that is comparable with plate glass.

Snorre Læssøe Stephensen

The Palm House at the Royal Botanic Gardens, Kew.

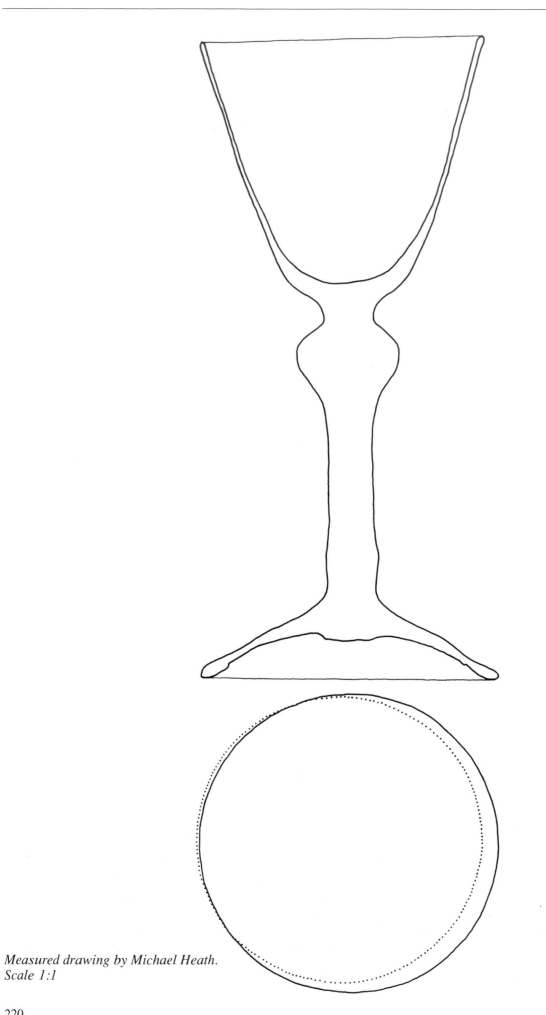

Measured drawing by Michael Heath.
Scale 1:1

Subject: Goblet, described in the original catalogue from 1763 as 'Vin Glas Chrystal'.
Manufacturer: Nøstetangen Krystal Glasverk, Norway.
Designer: From the times when designer and maker were one.
Manufacturing processes: Free blown bowl, stem and foot. The air spirals in the stem are made by pressing holes into the molten glass which are then covered with a new layer of glass and drawn out to form the stem. The rim of the foot is folded under to form a seam.
Dimensions: Height 17 cm.
Location: Aalborg Historiske Museum, Denmark.
References & notes: The *Nøstetangen catalogue* of 1763 has exceptionally good full-size illustrations of its products.
Related products: Aqua Vitae Glasses pp 226-227.

Evaluation: This is a drinking glass which would have been made to order in small series. Its distinguished character is the direct result of the glassblower's art and his design sense. It balances well when held, as he clearly intended, with thumb and first finger between the bowl and the bulge on the stem; the ample foot ensures stability for such a high glass, and the seamed edge to the foot strengthens the glass and makes for robustness in use – both characteristics of Norwegian glass of the period. The passage of time has brought another quality to the goblet which can only be appreciated by we who look back on it: it represents the early attempts of the craftsman to leave his functional tradition in order to serve the aristocratic world of style, and has that delightful character which occurs when the Baroque is interpreted with unsophisticated craftsmanship.

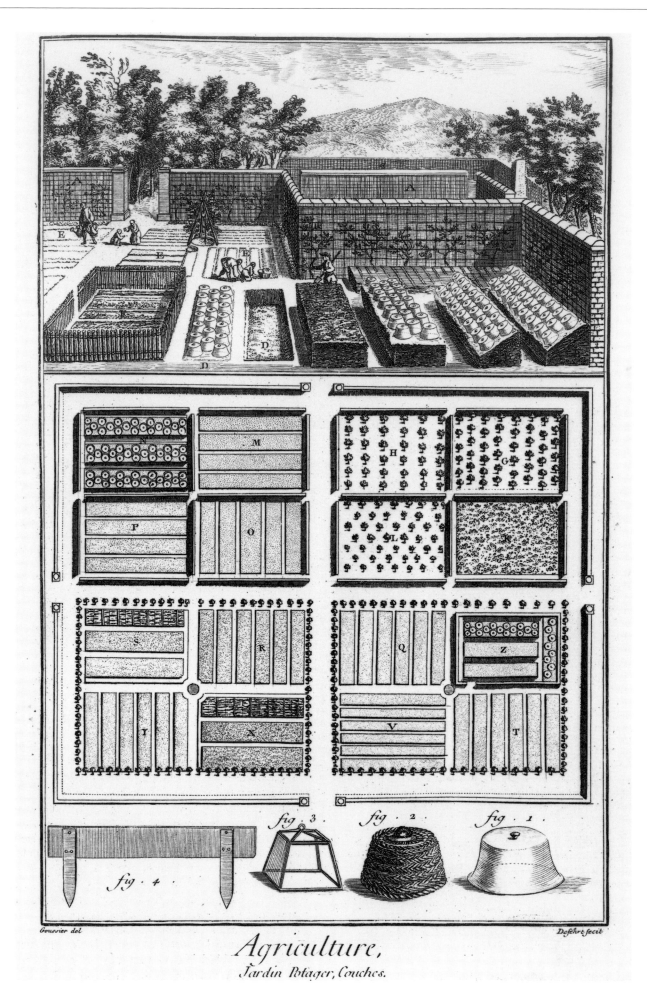

Agriculture,
Jardin Potager, Couches.

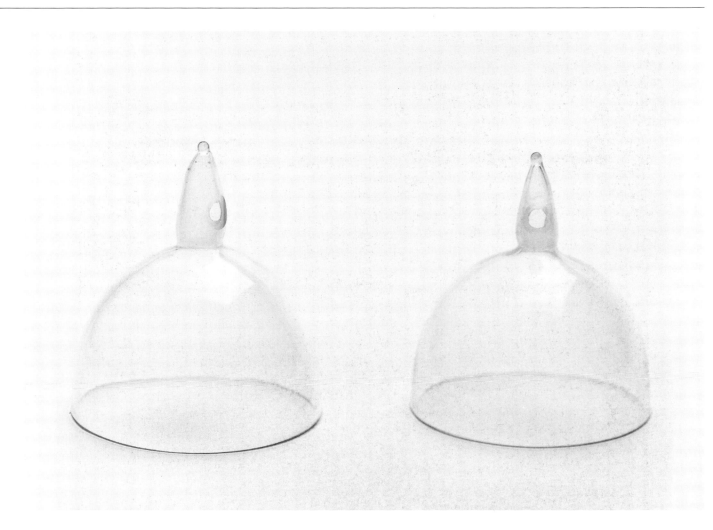

Subject: The engraving is from Diderot's *Encyclopaedia.* It shows two types of cloche (bottom) and cloches in use in an 18th century garden.

The photograph shows early 20th century cloches for seed germination.

Manufacturer: Fyens Glasværk, Denmark.

Manufacturing processes: Free blown, sprung and ground edge. The ventilation hole is ground.

Dimensions: Diameter 6.9 cm, height 7.5 cm, wall thickness 1.5-2 mm.

Location: Aalborg Historiske Museum, Denmark.

References & notes: *L'Enclyclopédie ou Dictionaire Raisonné des Sciences, des Arts et des Métiers* of Denis Diderot 1762-1777.

Fyens Glasværk's catalogue of 1910.

DLF-Trifolium A/S, seed and plant culture.

Evaluation: Here we have purposely juxtaposed the old and the new. The use of glass to create a mini climate in which seeds readily germinate, and young plants can develop independently of the cool spring weather outside, has a very long history. The cloche is an eotechnic invention and one which we now, some 300 years later, realise is of fundamental importance, it being one of the simplest ways we can utilise solar energy. Seed germinating cloches of glass are still used in plant cultivation laboratories in Denmark. Cloches of plastic have been tried but found to be unsuitable because they get scratched. The cloches shown here are very fine pieces of glasswork and resemble the present day ones, except that their ventilation holes are right at the top.

A condenser and candle in an engraver's workshop
in the Open Air Museum, Den Gamle By, Århus.

The historical development of artificial lighting has been, and still is, largely concerned with the conflict between light source and light treatment; or, light quantity and light quality. Ever since its existence glass has been at the centre of this struggle. The condenser and the lamp illustrate this in a rather special way.

The condenser is one of the many important eotechnical, or pre-industrial, inventions. It is, in effect, a large lens of water enclosed by a glass bulb. It was used by all sorts of craftsmen who had to be able to see the details of their work when the daylight was poor, or in the hours of darkness. It was a way of directing and concentrating light from an inadequate source, such as a candle or an oil lamp, on to the work. The water normally had spirit added to keep it clear. The condenser was hung on a string, or stood on a turned wooden stand ready to be placed between the light and the work.

The oil lamp shown here was a major step in the improvement of artificial light. It was designed by Aimé Argand in 1785 and its burner is literally the prototype of the final paraffin lamp development of over 100 years later. The strength of light from an oil lamp is dependent on the relationship between fuel type, wick size and air supply. The spherical oil tank is placed well above the burner so that the heavy plant oils, in use at the time, could be helped up the wick by gravity, having passed along the horizontal tube. The wick, in the form of a wide tape was led up into a cylindrical form in the space between a larger and a smaller diameter tube, in the same way as the best paraffin lamps of the early 20th century. The air supply to the round flame was up through the innermost tube. The air entered through the radially arranged slots below the burner. The glass chimney had three functions: to protect the flame from draughts, to make an upward current of air and to allow unrestricted light from the flame. The flame gave the light, just as it had done with the candle, on the basis that enough light was good light. It was not until some years later that opal glass shades were introduced which did much for the quality of the light. They distributed and reflected light, prevented glare and were in themselves a pleasing illuminated form.

Related products: Paraffin Lamp pp 246-247, and Desk Lamp pp 248-249.

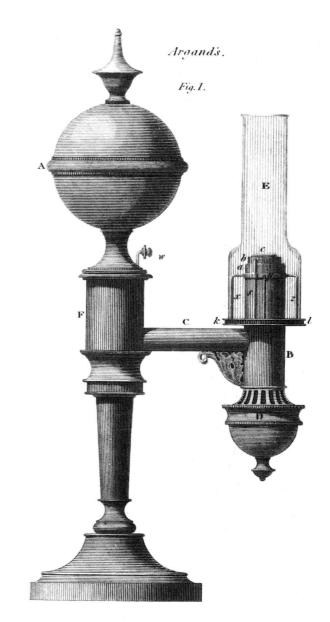

Argand's oil lamp from Rees's Cyclopaedia.

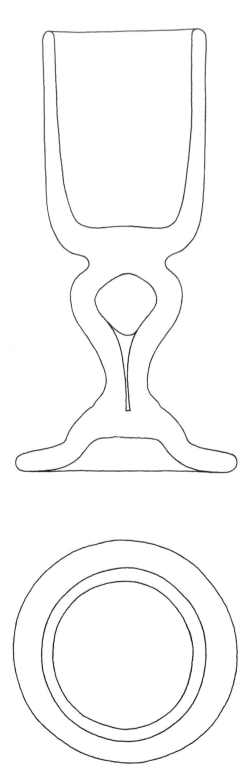

Measured drawing by Martin Bohøj. Scale 1:1

We have no details as to the production of these particular glasses but they represent a type which was produced in large numbers in Scandinavia during the 19th century. What we do know is that they were used for serving schnaps (aqua vitae) at an inn in central Copenhagen around the end of the 19th century. A similar glass is still made in Sweden. After the bowl is blown, the hole in the stem is made by pushing a wet wooden pin into the molten glass. The evaporation of the water in the hot glass blows it into a bubble. The absence of any attempt at the normal delicacy of the stem glass could indicate that they were a standard pattern for sale to public houses and restaurants. They hold 5 cl which is a very large measure by today's standards. Their robust, rounded, glassy character makes them good to drink from and they stand firmly on the table. They are a good example of a glassblower's standard product which, because no two glasses are quite the same, possess that delight of subtle variation peculiar to free blown glass. For related products see Goblet pp 220-221.

The three engravings epitomise early 19th century industry in England. The proud front facade; the works yard with workshops, stores and piles of fuel and raw materials and the canal with a barge being unloaded direct to the works; and production itself, here seen inside the glass cone with 20 men and children at work. The cone was the standard English glassmaking building from the 1730s to the late 19th century. It was the complete workshop in which all processes took place from the founding of the glass to the annealing of the finished product. The cone shape is derived from the centrally placed furnace, with its surrounding ring of pots of molten glass which gave ready access to many glassblowers at once. A chimney has, in effect, been made into a complete workshop in which a high temperature is maintained to avoid excessive heat loss from the glass on removal from the furnace. Air supply to the furnace came direct from the outside through underground flues. The outer ring of arches contained, amongst other things, the annealing hearths. The glass cone is a remarkable example of a building whose form is wholly decided by the optimum requirements of a hand manufacturing process.

Three views of Aston Flint Glass Works, Birmingham.

A dark green bottle of the period, Germany. Photo S.R. Gnamm, Die Neue Sammlung, Munich.

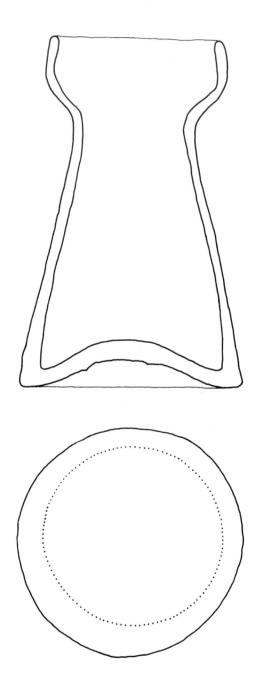

Measured drawing by Michael Heath. Scale 1:1

Subject: Glass for growing small bulb plants.
Manufacturer: Conradsminde Glasværk, DK.
Design: Traditional.
Materials: Lead glass with cobalt oxide added.
Manufacturing processes: Mould blown. The rim is cracked off and warm polished.
Dimensions: Height 9.5 cm.
Location: Aalborg Historiske Museum, Denmark.
References & notes: A similar shape is illustrated in a Norwegian glassworks catalogue from 1763. Larger glasses are made for larger species of bulb.

Evaluation: The practice of nurturing bulb plants into bloom, with the use of water, in a special cup-vase, originated in China and Japan and came to Europe in the 18th century. The Chinese used ceramic materials for the vessel but the European tradition has developed mainly in glass which gives the added advantage of showing the growing roots. This crocus glass is much more beautiful than our illustrations can convey. It is a robust, practical utensil whose soft, glassy character is enhanced by its deep cobalt blue colour, seen at its best on the window sill with the light behind.

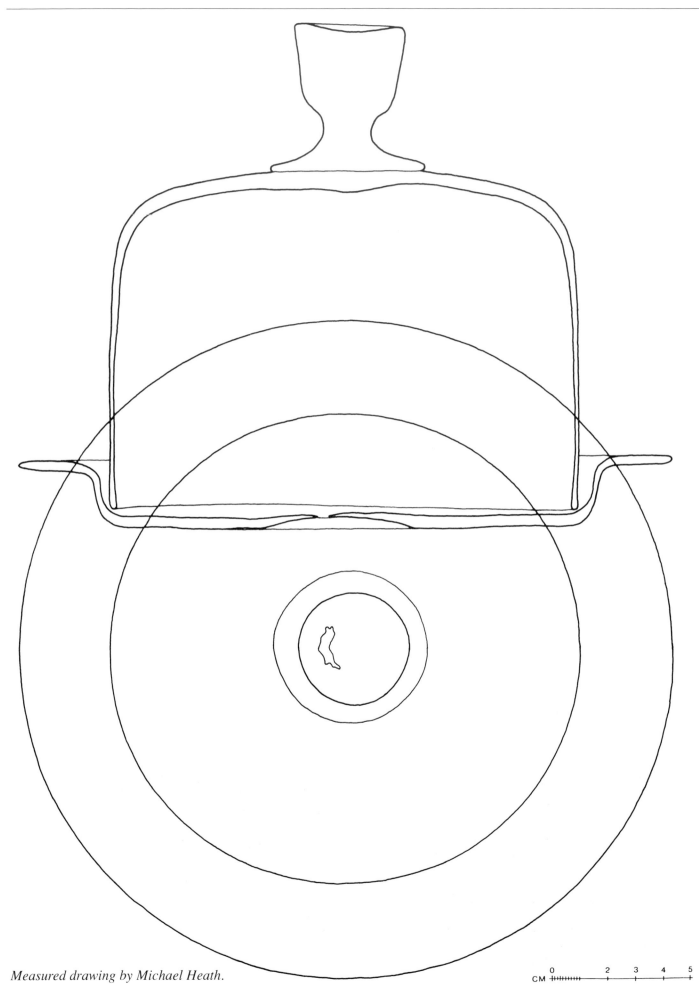

Measured drawing by Michael Heath.

Manufacturer: Probably Conradsminde Glasværk, Denmark.
Designer: Probably the glassworks.
Materials: Lead glass.
Manufacturing processes: The cover is free blown with trimmed side and edge. The solid handle is applied afterwards and is finished off by hollow grinding and polishing the top. The dish is also free blown with trimmed horizontal rim. The blowpipe break mark at the centre of the underside has been removed by hollow grinding.
Dimensions: The cover: diameter 17 cm, height 17.5 cm. The dish diameter 23.7 cm, height 2.2 cm.
Location: Aalborg Historiske Museum, Denmark.
References & notes: The dish has been used so much that the underside and edge are severely worn – to the extent of a hole appearing in the hollow grinding. This design of dish, or plate, was made in several sizes for various purposes.

Related products: The cloches from Diderot's *Encyclopaedia* pp 222-223.
Evaluation: This formidable piece of craftsmanship in free blown glass demonstrates the complete fitness of the material and the technique when applied to one of the simple utensils of the household. Apart from its main physical properties of cleanability, hardness, resistance to grease and smell, the glass and its form offer full view of the food – and an invitation to access – while at the same time afford full protection. The 'cheese bell', as the Danes call it, probably originated in France and it is interesting to compare it with the 18th century French cloche. There are Danish examples of the same bell being made and used for plant cultivation and cheese storage.

Subject: The Palm House at the Royal Botanic Gardens, Kew.
Builder: Richard Turner, engineer and contractor.
Glass manufactured by Robert Chance.
Designers: Decimus Burton and Richard Turner.
The glazing bars were a recent invention by
J.C. Loudon enabling construction of large glass areas.
Materials: Wrought iron, cast iron, glass.
Manufacture: An advanced system of prefabricated structural members bolted and riveted together on site. The original glass was probably cylinder glass. It was specially tinted green by the addition of copper oxide to reduce the heat of the sun. The slight, 8 m radius curve along the length of the panes was probably made by allowing the hot glass to cool on a curved mould.
Dimensions: Length of building 109.7 m. Height 19.8 m. About 16,000 panes of glass, normally measuring about 22 x 90 cm.
Location: Kew, nr. London.
References & notes: *Proceedings vol. 84, 1988,* The Institution of Civil Engineers, London.
The Greatest Glasshouse by Sue Minter, London 1990.
The survey measured drawings by the restoration engineers Posford Duvivier.
The Palm House underwent a major restoration in 1985-88 in which toughened glass on stainless steel glazing bars replaced the old materials.
Evaluation: Now, at the end of the 20th century, the Palm House at Kew remains one of the world's greatest examples of the harnessing of solar energy for the cultivation of plants. The timeless appearance of the building must be due to the overriding architectural idea of supporting glass in such a way as to allow maximum solar admission. The main structure and the glazing bars are slim but deep giving a very high ratio of glass to structure. The curved form gives a near right angle ray penetration and the building seems to be in accord with the great dome of the sun's passage overhead. The technology necessary to build this delicate, curved structure in iron was a complete innovation.

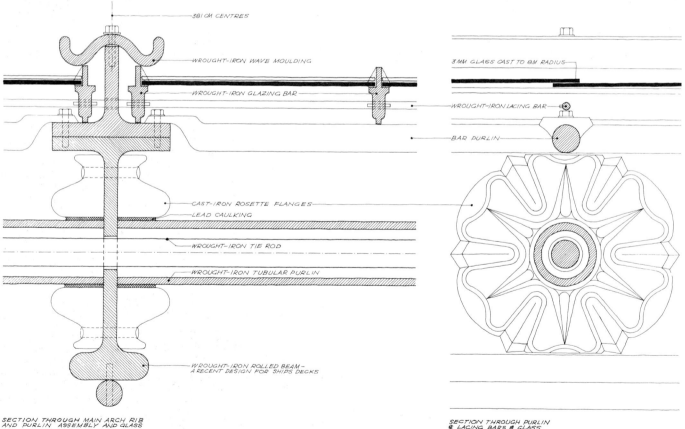

SECTION THROUGH MAIN ARCH RIB
AND PURLIN ASSEMBLY AND GLASS

SECTION THROUGH PURLIN
& LACING BARS & GLASS

Drawn by Charlotte Lemme after 1:1 survey drawing by Arthur Allen.

The Palm House seen from the South East.

Interior showing main arch ribs and purlins.

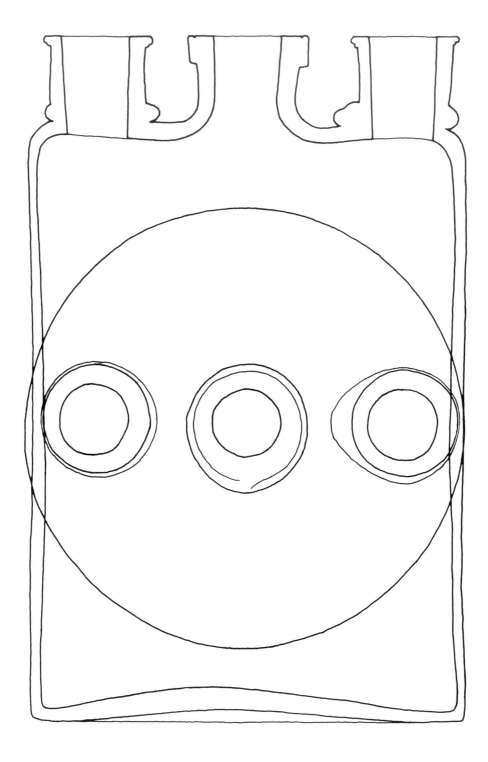

Measured drawing by Michael Heath. Scale 1:1

Subject: Part of an apparatus for the chemical cleansing of gasses.

Manufacturer: Holmegaard Glasværk, Denmark. First appeared in the catalogue of 1853.

Designer: The English chemist Peter Woulfe

Materials: Lead glass.

Manufacturing processes: The bottle, with its centrally placed neck, is free blown and trimmed. The two extra necks are melted on and all three are ground conically to receive rubber stoppers which hold glass tubes.

Dimensions: Diameter 11.4 cm, height 18 cm.

Location: Aalborg Historiske Museum, Denmark.

References & notes: *Chemie Für Laien,* by W.S.A. Zimmermann, Berlin 1858.

Related products: Laboratory glass pp 258-259.

Evaluation: These bottles are used in series in a row of between 4 and 6. They are connected by glass tubes which carry the gas to be treated from one bottle to the next, the gas passing through different chemical liquids which absorb the unwanted elements. The centre necks hold vertical glass tubes which also reach down to the liquids and act as safety valves. The robust proportions and weight of glass help to make the bottles stand firmly on the bench top. This apparatus is just one in a long line of historical inventions by scientists. They were usually necessary elements in a method of experiment or a laboratory process and were often named after their inventor. Glass was as often as not the ideal material for these tools of the laboratory.

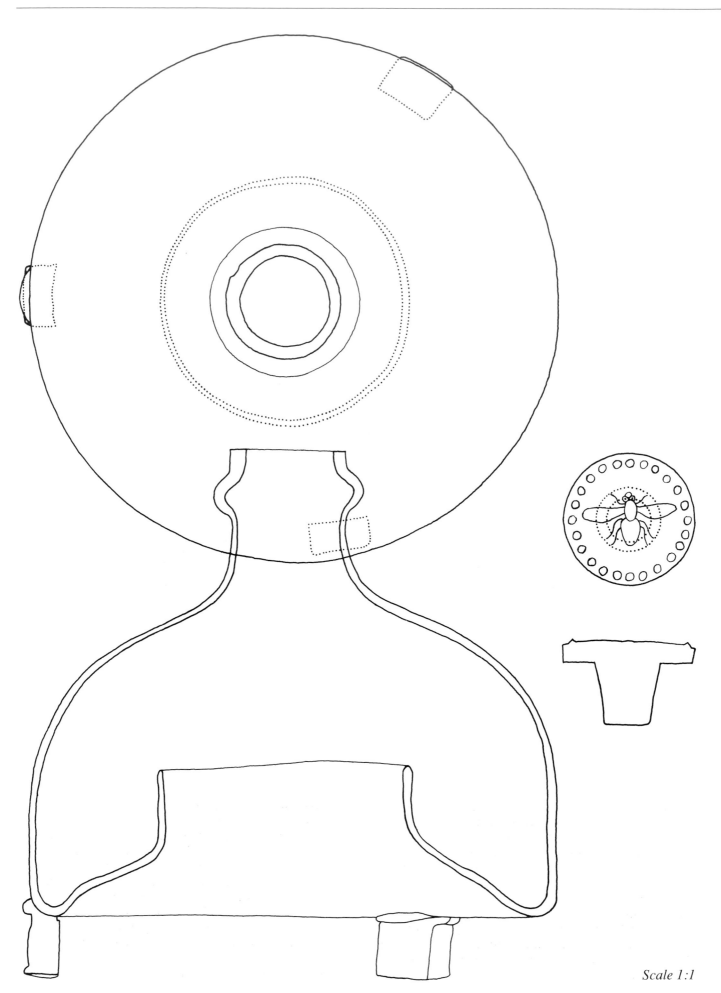

Scale 1:1

Subject: A Danish appliance for catching flies and wasps in the home. Sweetened water is poured into the 'moat' which attracts and drowns the insects.

Manufacturer: This example is made by Holme-gaard Glasværk in the early 20th century but the design is from the last half of the 19th century.

Materials: Lead glass.

Manufacturing Processes: The glass is free blown to a bubble. The bottom is cut away and pressed up into the bubble. The three feet are melted on to the hot glass. The stopper is made in a form and the fly motif stamped in the hot glass.

Dimensions: Diameter 14 cm.

Location: Private collection.

Evaluation: An interesting example of the use of the glassblowing skill to provide a piece of practical domestic equipment, the shape of which could not be produced in any other way. The resulting form seems quite natural to the material, the craft, and the function. When in use, however, this is somewhat marred by the spectacle of drowned and drowning insects – a fault which could be alleviated by the use of coloured glass.

Measured drawing by Michael Heath.

Photo: S.R. Gnamm, Die Neue Sammlung, Munich.

The bowls are English from about 1870. They are good examples of well-controlled free blowing and trimming. Here the break-off mark from the blow pipe at the centre of the bottom has been removed by hollow grinding, then polishing, in such a way as to give an added solid, prismatic quality to the glass.

The wine decanter has been in production in Denmark since the 1870s and is so to this day. Its height is 24 cm. It is made by the classic glassblowing technique: first the long throat is blown like a tube with plenty of molten glass kept at the bottom end; when the tube has cooled sufficiently to maintain its shape, the glass at the end is heated again and blown to its characteristic balloon form.

Blowing to the required shape like this, in stages, with the heating and cooling of the work, clearly illustrates the way free blown glass can be formed with such a high degree of control; the cold glass acts, in a sense, as a shaping tool by confining formability to the molten part.

We show these utensils together for the purpose of considering the timelessness which some glass objects possess and, while looking at them, to consider some of the less tangible yet unique qualities of the material, glass. It is so different from all the other materials that it is surprising that we take it absolutely for granted. Try for a moment to imagine you are seeing and handling a glass object for the first time. You will find that it has its own magic, and an infinite number of ever changing phenomena appear. The list is endless: clarity, 'clear as glass', transparency, it expresses its own form, all at once, all through, and the form and colour of whatever happens to be in or behind it. Glass is absolutely there, and yet it is not. It both reflects light and permits it to pass. It is as though this fluid material were in league with the fluid it often contains. It is at its best when the fluidity and movement of blowing are contained in the finished object. Blown glass and pressed glass are as two different materials, the latter lacking the vitality of the former. Glass is dependent on being clean – as soon as it gets dirty it becomes something else. Glass visually distorts that which lies beyond it, unless it is plate glass. Glass is the only material that casts more than just a shadow, it also casts light patterns. It pulls colours from surrounding objects into itself. The green nuances which are natural to the cheapest bottle glass are one of the material's greatest beauties. Glass is uncompromisingly hard. It is one of the few materials to which nothing has to be done after production to finish or protect it, it is self sufficient.

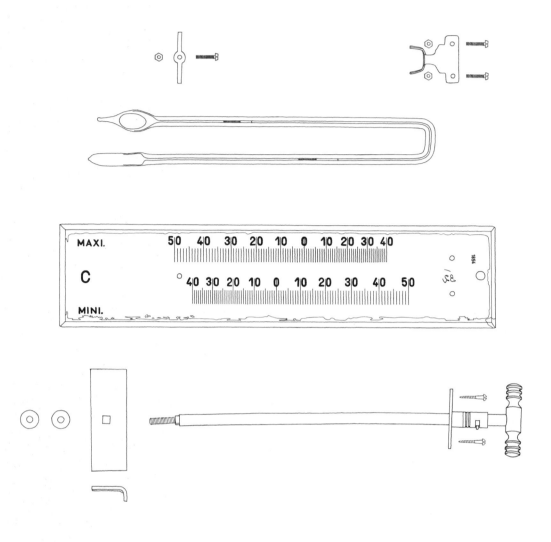

Subject: Thermometer for the measurement of outside air temperature.

Manufacturer: Geraberger Thermometerwerk, Germany.

Design: Works design.

Materials: Glass with fittings of German silver. Felt separates the fixing plate from direct contact with the glass. Mercury and spirit.

Manufacturing processes: The plate glass base is chamfered on all edges by grinding and polishing, figures and markings are engraved in the back face and the three fixing holes bored. Degrees below freezing point are painted red and the remainder black. Finally the white painted background is applied.

References & notes: This example is from the early 20th century but the type is earlier. The maximum-minimum thermometer was invented in the late 18th century.

Related products: Ebulioscope, pp 100-101.

Evaluation: The principal developments in temperature measurement, which were made necessary by industrialisation, took place in the first half of the 18th century with Ole Rømer's spirit thermometer in 1702, Fahrenheit's mercury thermometer in 1714, and Celsius' centigrade scale in 1743. Thus the modern thermometer shown here is the result of accumulated experience in which the glassblower and instrument maker Gabriel Daniel Fahrenheit played an important part. This type, which is operated from indoors, has been in extensive use in Northern Europe since the late 19th century and is still in production. The thermometer tube is, of necessity, glass. Because this material, apart from being transparent, is ideally suited to withstand all weather conditions, is repellent to dirt and is easy to clean, glass has been chosen for the entire instrument. Thorough cleaning, which is seldom necessary, is easily done by dismantling the whole assembly. At the time of writing this thermometer has been in permanent use for over 30 years.

Measured drawing by Martin Bohøj. Scale 1:2

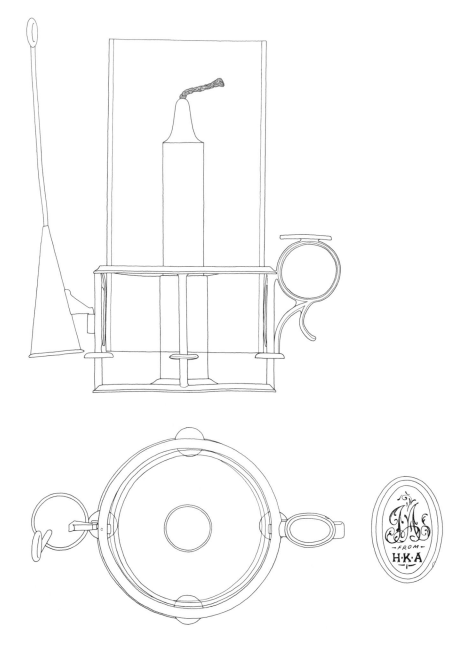

Measured drawing by Martin Bohøj. Scale 1:2

Engraving on thumb-plate. Scale 1:1

Subject: Portable candlestick.
Manufacturer: Elkington & Co., Birmingham. Probably made in the 1890s.
Designer: Company designer.
Materials: German silver (alloy of copper, zinc and nickel) electroplated with silver. Glass shade.
Manufacturing processes: Silversmithing. Machine assisted handwork. Largely made of bent and soldered pre-rolled sections. Electroplated with silver after assembly. Polishing. Glass, mould blown. Edges cracked off and ground.
Dimensions: Height to top of glass 18.5 cm, height of glass holder 6.7 cm, outer diameter of glass holder 9.5 cm.
Location: Author's collection.
Evaluation: In 1836 the makers of this candlestick

G.R. and H. Elkington started taking out patents concerning the electroplating of copper, and its alloys, with silver and gold. In 1840 they started the manufacture of electroplated articles, and this candlestick was one of their catalogue designs. Bearing in mind its date in the middle of the Victorian period it is surprisingly simple and uncluttered. It gives a good and unobstructed light right down to the surface on which it stands, and the flame is almost invisibly protected by the clear glass cylinder. The glass is removable for cleaning, but when in position is gently held from rattling by the flat spring soldered to the column, which also supports the snuffer. The handle and the holder's thumb and fingers 'melt' together in a positive grip. Everything has its clear place and function. When there was no such thing as the electric torch this portable light must have been a prized possession.

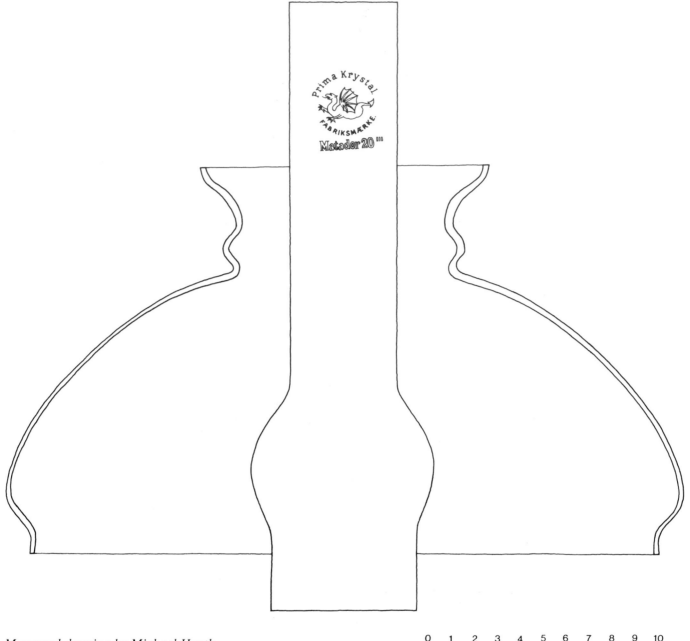

Measured drawing by Michael Heath

CM 0 1 2 3 4 5 6 7 8 9 10

Subject: Hanging lamp for many uses but typically for lighting the dining or work table.

Manufacturers: Chimney glass, Fyens Glas-værk, Denmark. Shade, Czechoslovakian. Tank, probably Danish. Burner, Erich und Graetz, Berlin. Frame, Svend Arnkilde, Copenhagen.

Materials: Chimney, clear glass. Shade, opal glass. Tank, brass. Burner, brass with cotton wick. Frame, mild steel, painted black.

Manufacturing processes: Chimney, made as a tube. The lower end of the tube re-melted and opened and trimmed to the required size of burner. Trademark transferred to glass and fired. Shade, free blown to a template. Top and bottom edges sawn with a diamond cutter (edges used to be folded which was a stronger and more pleasing solution).

Dimensions: Frame, height 53 cm, greatest width 39 cm, shade ring diameter 29 cm.

Location: Author's collection.

References & notes: The 1924 *catalogue of Kastrup Glasværk,* Denmark.

The five parts of the lamp have different production dates but the design of each, with the probable exception of the frame, are from the first decade of the century, or earlier.

Related products: Argand's Oil Lamp pp 224-225, Desk lamps pp 248-249.

Evaluation: With the discovery of paraffin the shape of the oil lamp changed, its lighting effect increased and it burned cleaner and with less smell. The change of shape was radical: the heavy vegetable oils used in earlier lamps had to be gravity fed to the wick, which meant that the tank had to be above and to one side of the burner, always causing an unwanted shadow. The thinner paraffin oil could rise in the wick by capillary action alone with its tank placed neatly under the burner and the wick going straight down into it. As can be seen from the list of manufacturers, the paraffin lamp was frequently composed of components from several different factories. These were available in various standard sizes which were assembled at the lamp shop in different combinations according to the customer's requirements. The glass factories made all lamp parts, except the burners and frames, including a tank shape like the one shown here. The paraffin lamp is an early, and quite elaborate, example of the necessity of standardisation.

The opal glass shades were often well shaped, as here, and together with the soft light of the flame, gave a beautiful illumination of the room.

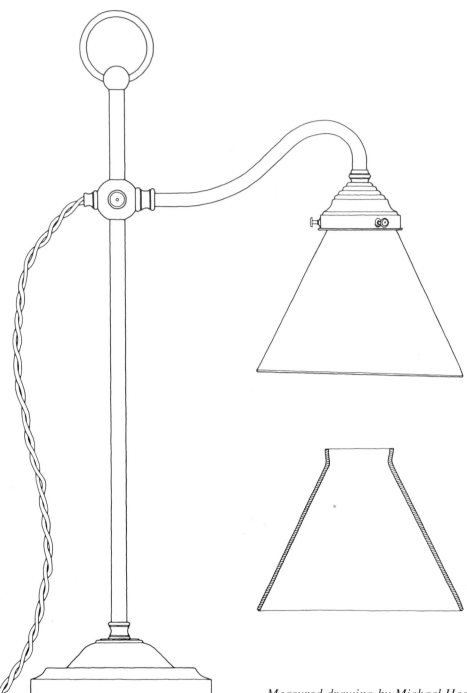

Measured drawing by Michael Heath. Scale 1:3

Subject: Electric desk lamp with adjustable height. Also shown is an oil lamp with Argand-type burner and gravity feed, originally designed for the thicker, pre-paraffin vegetable oils.

Materials: Brass and glass with cast iron weight in the base to keep the lamp standing firmly. Porcelain and copper bulb socket.

Manufacturing processes. Metal parts: spinning, casting, turning, threading. Glass shade: opal glass with external flash of green pigmented glass, free blown, trimmed to cone form.

Dimensions: Total height 55 cm.

Location: Private collection.

Related products: Condenser and lamp pp 224-225, Paraffin lamp pp 246-247, Light bulb pp 50-51.

Evaluation: The two lamps shown in the photo indicate that the production of electric lamps slid in quite naturally beside a continuing paraffin lamp production and adopted many of the same design details, a typical example of the overlapping which occurs all through industrial development. The electric lamp's glass shade, composed of green glass outside to reduce the glare and white glass inside to increase reflected light, is one of those basically good solutions which, with the brass stand, had its own special functional and visual quality. First with oil or paraffin, later with gas and finally with electricity the green glow of this composite glass shade characterised the office interior over a long period.

Photo. Louis Schnakenburg

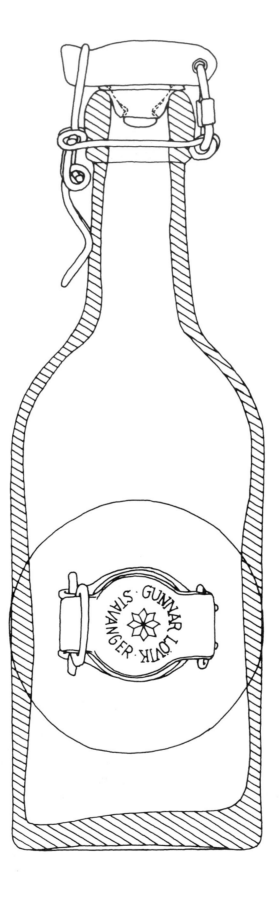

Measured drawing by Claus Bech-Danielsen. Scale 1:1

Subject: Mineral water bottles for repeated distribution.
Manufacturer: The drawn example, unknown, Norway.
Photo: left, Årnäs, Sweden. Centre, unknown, Denmark.
Right, unknown.
Designer: Works designs. Lever, or swing stopper
patented by Friedrik Siemens 1880s.
Materials: Soda glass, technical porcelain, 2 mm tinplated steel wire. Rubber washer.
Manufacturing processes: 4 different factories involved
in making each bottle: Bottles. mould-blown in semi-
automatic bottle machines. Stoppers, pressed in moulds.
Two have printed trademarks in chrome oxide (green)
under the transparent porcelain glaze. Fired to sintering,
1200-1300°C. Swing (or lever) fittings: 5 pieces of 2 mm
steel wire formed in jigs and fixed to the bottle necks
with crimped locking strip.
Weights: The bottle in the drawing 475 g. Its liquid
contents 300 g (compared with present day bottle/content
ratio of 250 g: 250 g).
Location: Author's collection.
References & notes: *Danske Flasker fra Renæssancen
til vore dage,* by Mogens Schlüter, Copenhagen 1984.
Related products: Milk bottles pp 256-257.
Evaluation: The glass bottle is older than the glass

windowpane and it is these two functions that have been
by far the most important for the glass industry. Since
glass for windowpanes stopped being made by the glassblower, and became sheet glass, the industries have been
quite separate and everything about them is different,
except the basic raw materials. The bottle, with its vast
number of different shapes for different purposes, was so
important to society that it became a locomotive in the
development of mechanisation. These four bottles are
typical products of the early consumer society: mass-
produced but composed of semi-handmade and semi-
automatically made parts. Although it bears signs of
having been designed by different people at different
times and places, the bottle with swing stopper is an
ingenious solution to a difficult problem: an almost
hermetic, self-contained bottle and stopper that can withstand the pressure from a gas-charged beverage and that
can be opened and closed and returned for re-filling and
re-sale. Emballage and utensil in one. The use of the
three hard-wearing materials, each with their clearly
defined functions, is very satisfactory. One remembers it
from childhood, though, as being a rather alarming thing
for small fingers to deal with.

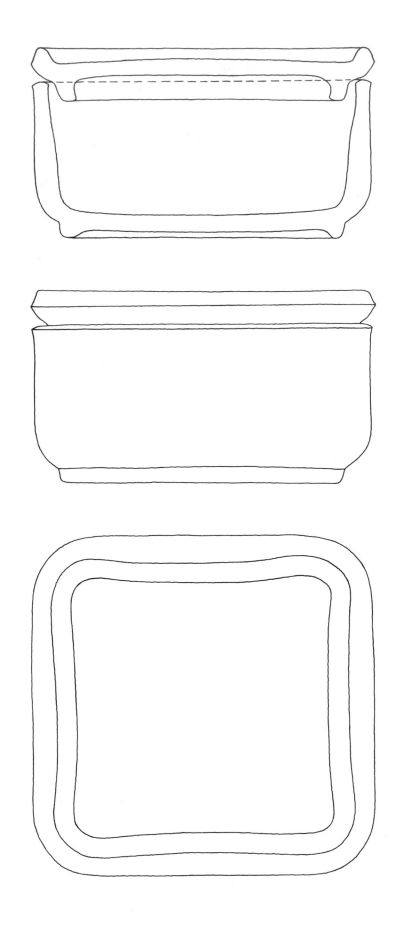

The smallest container of the set with lid. Measured drawing by Martin Bohøj. Scale 1:1

Photo: S.R. Gnamm, Die Neue Sammlung, Munich.

Subject: 'Kubus' kitchen storage boxes, 1938.
Manufacturer: Lausitzer Glasverein, Germany.
Designer: Wilhelm Wagenfeld.
Materials: Soda glass.
Manufacturing processes: Pressed glass process in which the molten glass is rapidly pressed into a mould by a plunger.
Dimensions: There are three heights of container: 16.3, 8.3 and 4.3 cm, including location bead beneath. Three plan dimensions: 18 x 18, 18 x 9 and 9 x 9 cm. The three sizes of lid are 1.4 cm in height, including location bead.
References & notes: *Industrial Design,* by John Heskett, London 1980.
Evaluation: We have chosen this well-known example of early domestic functionalism, not only for its design qualities, but also because it is a very good example of pressed glass. It was designed for the storage of foods in the refrigerator, when this was just coming into domestic use in Europe. It is well-considered in every way, not least in its dimensioning which is very satisfactory in use. In this connection the 'cubic' stack, as shown, is misleading; the containers do, in fact, function well individually, though it is useful to be able to put one unit on top of another with or without lids. The pressed glass process permits a constant wall thickness with comparatively little distortion and dimensional tolerance. Location beads on the undersides of lids and containers are formed so that they locate at corners only. The inevitable mould parting ridge round the edges of lids is placed so that it serves to improve finger grip when lifting. The transparency and shine of free blown and mould blown glass is lost in pressed glass.

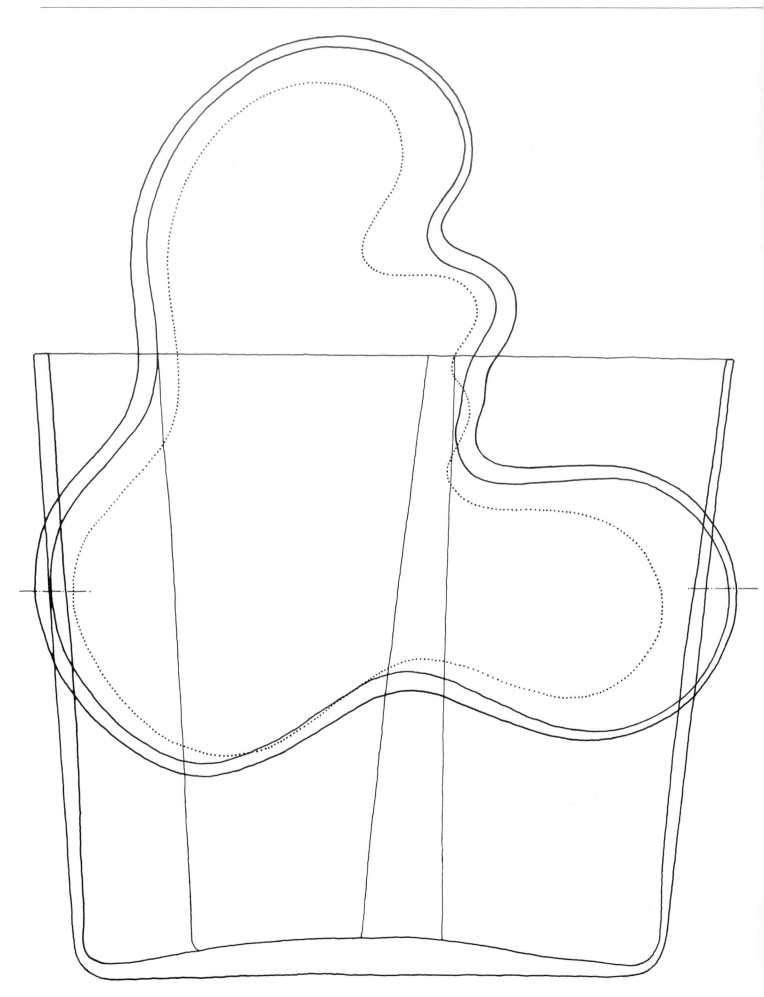

Subject: Known as the Savoy vase, this design is based on the sketch design for a competition promoted by Karhula Glassworks in 1936
Manufacturer: Karhula Glassworks 1936-1949, Iittala Glassworks 1949 onwards.
Designer: Alvar Aalto.
Materials: Crystal glass.
Manufacturing processes: Blown to a wooden mould before 1954, to cast iron moulds since. Edge sawn with diamond cutter and polished.
Dimensions: The vase shown is 165 mm high. Through the years larger and smaller sizes have been made.
Location: Author's collection.
References & notes: *Alvar and Aino Aalto as Glass Designers,* by Satu Grönstrand, Iittala– Nuutajärvi Oy, Finland 1988.

Evaluation: Alvar Aalto's vase has an idea and a form which nobody ever seems to get tired of. The vase looks as natural as the flowers and the foliage that it holds. Even the plant stems, and the water they are in, are allowed to take part in this almost magical meeting between plant and vase. Because of the vase's special facility for allowing spread flower arrangements, the clusters of the herbaceous border can be 'transplanted' indoors. The fluid, meandering form is explained and defined by a precise, horizontal conclusion – the ground and polished glass edge which, with its fluctuating thickness and curves, tells the inner story of the form beneath. The bottom of the vase is formed so that it stands, without rocking, on three points.

Measured drawing by Michael Heath. Scale 1:1

A

Measured drawing by Martin Bohøj. Scale 1:1

B

A *Danish Standard ½ l.*
1930s–1960s

B *English dairy-group*
1 pint. 1990s

C *Danish brown ½ l.*
1959–1970

Subject: Three milk bottles compared.
Manufacturers: Holmegaard Glasværk a/s, Denmark.
Rockware Glass Ltd, England.
Designers: The brown bottle is by the Norwegian
engineer Hougen. The other two are works designs.
Materials: Bottles **A** & **B** sand, soda ash, lime-stone and
cullet. Bottle **C** has iron oxide added to reduce the effect
of sunlight on the milk.
Manufacturing processes: Mixing of ingredients,
melting to a temperature of about 1600°C, passing to
forehearth and cooling to about 1200°C before draining
through orifice. At this point the material is cut to exactly
the right weight. These gobs then go to the first mould
which forms the bottle's mouth piece, then on to the
second mould which blows the still molten glass to the
finished shape. The bottles pass through the lehr. Most of
these processes are automatic.
Weights and volumes: Bottle **A** 415 g, 500 ml;
bottle **B** 235 g, 568 ml; bottle **C** 365 g, 500 ml.
Related products: Lever top bottles pp 250-251.

Evaluation: When in use the milk bottle is the principal
performer in a programme of operations carried out both
by machine and by hand. It starts its working day within
the automation of the dairy and ends it in somebody's
kitchen in the role of a personal utensil. After use the
bottle is returned to the dairy and the whole procedure
starts again and is repeated up to about 20 times. Finally,
it is crushed to become cullet for remaking. Most of the
inherent qualities of glass, not least that of hygiene, make
the material ideally suited to this strenuous programme.
The exceptions are its brittleness and its weight. Glass
technology and design have concentrated on these
deficiencies as can be seen by comparing the weights and
wall thicknesses of bottles A and B. Bottle B has a ratio
of height to diameter which is ideal for glassmaking. The
manufacturers maintain that this and the curved form of
shoulder and heel have enabled the forming process to be
carried out with a considerably reduced wall thickness,
while maintaining strength. The reduction of weight was
important in England until recently when milk was delivered to the home by the milkman in his electric van.

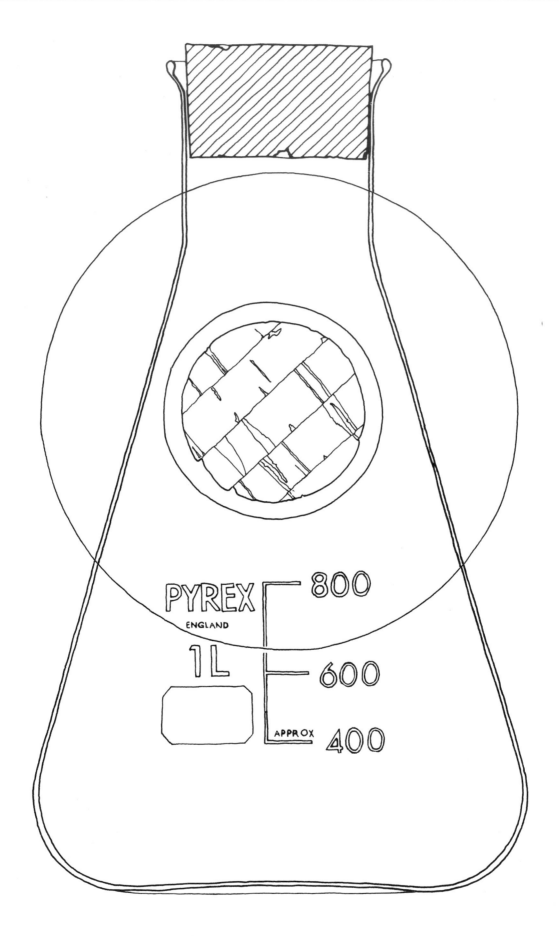

Measured drawing by Claus Bech-Danielsen. Scale 1:1

Subject: Two Erlenmeyer swirling flasks and a crystalli-sation dish.

Manufacturer: Bibby Sterilin Ltd, England.

Designer: The flasks are named after their designer, the chemist D.Erlenmeyer (d.1909).

Materials: Borosilicate glass.

Manufacturing processes: We have been unable to obtain information applying specially to this type of glass, but it is composed of varying quantities of silica, alumina, calcium, sodium oxide, and boric oxide. Labels, printed and fired.

Dimensions: Small flask, height 10.4 cm. Large flask, height 22 cm. Dish, height 8.8 cm, diameter 17cm.

Location: Author's collection.

References & notes: *The Encyclopedia of Glass,* by Phoebe Philips, 1987.

The form of these vessels has hardly changed since the 1920s. The inscription on the flasks shows that these examples are later than the 1960s.

Related products: Tea Set, see following page.

Evaluation: These vessels have one of the longest production periods of any in this collection. Because of the material of which they are made, they are good examples of a development within science and industry which has spread to other spheres of life. Borosilicate glass expands only about one third as much as common glass when heated, and as a result it is less apt to break under rapid temperature change. It is also more resistant to chemicals and is an excellent electrical insulator. These properties have made the material well-suited for ovenware and other heat resisting utensils, for electric light bulbs, thermometer tubes and thermos flasks – just to name some of the most common uses. Blown articles, such as the ones shown here, possess a special quality of delicate lightness, which is exploited very well in certain domestic utensils. It is due to the strength of the borosilicate material, which allows thinness, and constant wall thickness – both of which make for heat resistance.

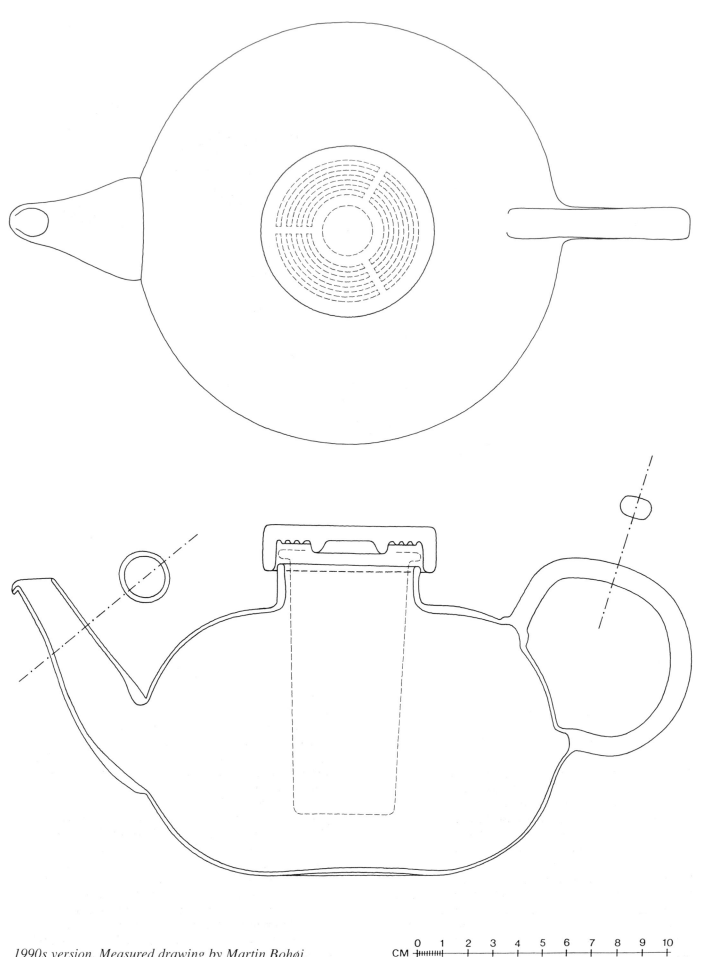

1990s version. Measured drawing by Martin Bohøj.

CM 0 1 2 3 4 5 6 7 8 9 10

Photo: S.R.Gnamm, Die Neue Sammlung, Munich.

Chronology: The photograph shows the original tea set manufactured by Jenaer Glaswerke Schott & Gen. from 1932. The teapot shown in the measured drawing is a revival of the original design marketed by Bodum in the 1990s.

Designer: Wilhelm Wagenfeld.

Materials: Borosilicate glass.

Manufacturing processes: The pot is mould blown. The strainer and lid are pressed. The spouts and handles are free formed,which accounts for the variations in shape between those of the 1930s and the 1990s.

Dimensions: The example shown in the drawing: length (spout to handle) 27.2 cm, height 14 cm. Capacity 1.5 *l*. Weight 400 g

Location: The tea set is in Die Neue Sammlung, Staatliches Museum für Angewandte Kunst, München. The Bodum example is in the authors' collection.

References & notes: The tea strainer which is shown with a dotted line on the drawing is not original, being plastic and of different design.

Related products: Laboratory Glass pp 258-259, Teapots pp 180-181 and 260-261.

Evaluation: The light, clear quality of blown borosilicate glass is shown at its best in this tea set. This type of glass was invented primarily for industrial use, and for the laboratory, as a heat- and chemical-resisting material. Wagenfeld was one of the first designers to exploit it for domestic use. Its comparative strength permits thinness and therefore lightness. This 1.5 litre teapot weighs only half as much as a 1 litre traditional earthenware one. Lightness is important in a teapot which is normally held with one hand only, often at arm's length. The form of the handle enables a good grip and control, largely due to the straight thumb piece, and the borosilicate glass is a good insulator so the handle does not get too hot to hold. The spout pours well if the glassmaker has been success-ful in making exactly the right form. The spout shown in the drawing pours without turbulence or dripping. This tea set has its own inherent 'decoration' – the beautiful colour of the tea seen through the glass.

Subject: 'Caravelle' glasses, for general use.
Manufacture and design: VMC, France.
Materials: Soda glass.
Manufacturing processes: The glasses are machine made in steel moulds. The rims are cracked off and warm polished.
Capacities: 2.5 cl, 4.5 cl, 10 cl, 20 cl, 25 cl. The series also includes a high 29 cl tumbler.
Location: Author's collection.
Evaluation: The secret to the quality of these simple glasses – apart from their utter simplicity – is best discovered by looking at their sections. The thickness and form of the glass at the bottom gives the glasses many qualities which the user feels as much as he sees. It gives weight at the bottom which makes them stand well and balance well in the hand. It gives a round bottom to the inner glass which makes it easy to clean. It exploits the material's prismatic play with light and colour. The glasses are sufficiently conical to make them secure to hold, easy to pour out the last drop and easy to dry. The glasses look well-proportioned in all sizes.

Measured drawing of the 2.5 cl and 25 cl sizes by Michael Heath. Scale 1:1

GLOSSARY

ABS Acrylonitrile-Butadiene-Styrene.

ACETYL RESIN Polyacetat.

ADZ Type of axe, but with the blade set at right angles to the handle. The chair-maker's adz is called a 'howel'.

ANNEAL (metal and glass)
The heating, and controlled cooling process which relieves a hot (though not molten) material of its inner tensions.

BALL CLAY Secondary, sedimentary clays. The name derives from the large balls into which it was made for horse transport.

BALUSTER The upright support to a handrail, or chair arm.

BAND SANDER (wood)
Machine on which the sandpaper is in the form of a continuous band moving over a table.

BEADING A rolled groove in thin metal, which on the reverse side is a linear projection. It has the effect of stiffening the thin material, and can also be a form of decoration.

BEVEL To cut or grind to an angle, as in the cutting edge of a chisel, or axe, or the measuring edge of a ruler.

BLOCKBOARD Wooden sheet material made in large panels made up of strips of wood, glued side by side, and enclosed on both faces by veneer. The veneer is glued to the strips under pressure, with its grain at right angles to them, to form a flat, relatively stable, panel. Various thicknesses.

BOLE The lower part of the trunk of a tree, which is free from branches. The part which is sawn to planks.

BOROSILICATE GLASS A thermal shock-resistant glass consisting of silica, alumina, calcium, sodium oxide and boric oxide.

CHIPBOARD or PARTICLE BOARD
Fine wood chips mixed with glue and rolled out under pressure to form large flat panels. Various thicknesses.

CONVERT (timber) To saw the tree trunk, or bole, into straight planks, beams, etc., at the sawmill.

CORNISH STONE also CHINA STONE A feldspathic mineral.

CRAMP (or CLAMP) A screw device for holding the parts of the work firmly together while making or glueing

CULLET Recycled, crushed glass.

CYLINDER GLASS An early method of making window glass whereby the glass was blown to a cylindrical form. The hot glass was then clipped along the length of the cylinder and laid out to cool on a flat mould.

DIE Tool for making the thread on a screw or bolt. (The tool for making the internal, female, thread is known as a 'tap'). Or, the mould which gives a work piece shape in stamping or forging. Or, the orifice through which metal is forced to become an extrusion.

DISC SANDER The sanding machine in which the sandpaper is fixed to a disc.

DOUBLE LAP SEAMING Jointing tin plate etc. by the folding together of, for example, the bottom and sides of a 'tin'. Originally done by hand with a special seaming tool. Later with a small hand-turned rolling machine. Now automatic.

EOTECHNIC The phase in the history of technology of water and wind power, and natural materials worked by hand.

ERGONOMICS The design aspect of the relationship between human beings and their working environment.

FAIRING Shaping the edge of a wooden strip or plank so that it is narrower at the ends.

FELDSPAR A large group of minerals which have decomposed from granite and igneous rocks and are thus allied to clay.

FENCE A guard or guide, as for regulating the movement of a tool or work.

FLUSH JOINT Where the surface, or surfaces, of two jointed components are flush with each other.

FORGING Forming metals by use of heat and impact, with or without dies. Sometimes performed cold.

FREE BLOWN Glass objects made by the glassblower blowing through the blow iron into the molten glass at the opposite end, and the other movements necessary to achieve the desired shape. (see 'trim')

FRIT Very clear glaze composed of various highly refined minerals. Similar to glass.

GALVANISE The application of a thin coat of zinc to iron or steel, by dipping the object into molten zinc, or by electrolysis, to protect it against corrosion.

GERMAN SILVER An alloy of copper, zinc and nickel. Also known as 'new silver'.

GLOST The glaze firing of ceramics.

HARDENING The heating and sudden cooling (quenching) of steel alloys, which increases their strength and hardness.

HELICAL FLUME CASING A water conducting tubular casing with interior formed like a screw thread.

INCH
One inch (") = approx. 2.55cm;
one foot (') = 12 inches = approx. 30.50cm
one yard (yd) = 3 feet = approx. 91.4cm.

JIG A device for steering the work while machining. The jig ensures that a number of components are identical.

JIGGERING (ceramics) An expression derived from the 'jigger' which is a pivoted arm for holding a template. Used for turning the interior and rim of hollow ware, while leather-hard, and still in the mould.

KAOLIN China clay.

LAMINATE To glue together layers of wood under pressure, usually to achieve curved forms of great strength, stiffness,

and/or resilience. Narrow laminations have single grain direction; shell-like laminations have crossed grain.

LEAD GLASS Also called 'crystal' glass, having a high content of lead monoxide which makes the molten glass easier to work, and gives the finished product great clarity, a fine surface and softness of form.

LEATHER-HARD (ceramics) The stage in the drying of clay when it becomes almost rigid, but still moist.

LEHR A long tunnel-shaped oven for annealing glass.

LUMBAR The small of the back. That part of the human body for which it is important to give adequate support in chair design.

MALLEABLE The ability of a metal to be formed by impact, pressure, or rolling, without fracturing. Ductile.

MORTISE AND TENON One of the most important joints in carpentry, joinery, furniture making etc. The mortice is a hole in the side of a piece of wood, the tenon is a projection from the end of a piece of wood, which is cut to exactly fit the mortice. The joint is usually secured with glue (see 'shoulder').

MOVEMENT The works of a clock.

NACELLE The enclosure which houses the engines of an aeroplane, or the machinery of a wind turbine.

NEOTECHNIC The phase, in the history of technology, of electricity, the internal combustion engine, alloys and plastics.

OGEE A section, often a moulding, consisting of adjacent convex and concave curves.

PALAEOTECHNIC The phase, in the history of technology, of steam, coal and iron, and the beginnings of mass production.

PBTP (or PBT) Polybutyleneterephtalate.

PIN JOINT The connection of two members with a single screw or bolt. The joint can act as a hinge unless the two members are secured elsewhere.

PLYWOOD Veneers glued and pressed together with their grain direction alternating at right angles to each other. Plywood is made in large flat panels of relative stability. Various thicknesses.

PRESSING The normal method used for shaping thin sheet metal to convert it from flat to three dimentional. Apart from achieving a required form, pressing is often done to stiffen. Also: the commonest industrial method of forming molten glass to make it into poducts.

P.S.I. Pounds per square inch.

PUNCH A humorous periodical published in Britain.

QUARTER SAWN (wood) Sawn at right angles to the tangent of the annual rings, i.e. so that the portion of the annual rings which show at the end of a plank etc., go across, in the shortest direction, from face

to face – or nearly so. Because of the way wood shrinks when it dries, this makes for the most stable and rectangular sectioned plank, with a narrow, straight grain on its two faces. In some woods medullary rays are exposed, giving a characteristic figure.
RATTAN A climbing palm from S.E. Asia. Its outer fibres are peeled off in flat, narrow strips and are typically used for 'honeycomb' caning of chair seats and backs. The inner fibres are made into round sectioned strips, also for caning.
RIP SAW (wood) To saw with the grain. A saw with teeth designed to do this.
R.P.M. Revolutions per minute.
SAGGAR A clay box in which pottery is fired to protect it against flame and ash.
SCANTLING Timber with small cross section dimensions.
SCARF JOINT Joint used to join two pieces of wood end to end, in the form of an angled overlap.
SHOULDERED JOINT (wood) Joint with a tenon cut with 'shoulders', normally on two or four sides. The shoulders butt on to the wood surrounding the mortise.
SLIP Clay mixed with water to a smooth,

creamy consistency.
SODA GLASS Glass named after its soda flux which makes the sand or flint melt and become glass.
SPINDLE MOULDER One of the principal machines in woodworking. A fast rotating, vertical spindle, projecting up through a cast iron table, with a cutter head onto which profiled cutters are fixed, so that components can be edge or face moulded.
SPINNING OF METAL To shape into hollow, rounded form, thin sheet metal during rapid rotation on a lathe-like machine. The sheet metal disc to be formed is gradually pressed against a wooden pattern of the required shape with a special tool. This is a very good example of the malleability of certain metals.
SPRIGGING Joining more clay onto the main part of a pot, such as a handle, or added decoration (sprig). The surfaces to be joined have to be softened with water or slip.
STRETCHERS The struts of a chair which 'stretch' from leg to leg, or between stretchers (called cross stretcher) and are part of the underframe of the chair.

TAPERING Grinding the sides of a knife blade to make it gradually thinner from handle to blade end.
TEMPER To make hardened steel less brittle by heating it to a temperature which removes the inner tensions which have occurred under hardening.
TEMPLATE A pattern, or guide, usually in the form of a thin 'plate', made for repeated use to guide the shaping of the parts of, for example, a chair.
THICKNESSING Planing a board to a predetermined thickness.
TRANSFER (ceramics) The method whereby ready-designed, two-dimensional, patterns etc. can be transferred to pottery, and fired, to become a permanent part of the glaze.
TREADLE Foot pedal.
TRIM To control the form of freeblown glass by the use of wetted wooden tools etc. applied directly to the hot glass.
VORTEX A whirling mass of water - as in a whirlpool.
YAW The motion of a boat, aeroplane or wind turbine about its vertical axis.

MANUFACTURERS whose products are represented.
Sixty percent of these exist in 1999, although sometimes under different names.

Austria
Gebrüder Thonet

Denmark
Aksel Skov
Aluminia
Bodum
Carl M. Cohr
Conradsminde Glasværk
Erik Mangor
Fritz Hansen
Fyens Glasværk
Holmegaard Glasværk
J.P.B.O.
Kastrup Glasværk
Louis Poulsen
Munch Møbler
N.E.G Micon
Porclæns Fabrikken Norden
P.P. Møbler
H. Rasmussen & Co.
Royal Copenhagen Porcelain
Rømer's Tobakspibefabrik
Scandia
Scanwood
Snorre Læssøe Stephenson
Svend Arnkilde

Finland
Fiskars
Karhula Glassworks
Iittala Glass

France
VMC

Germany
A.E.G.
Erich und Graetz
Geraberger Thermometerwerk
Jenaer Glaswerke
Lausitzer Glasverein
R. Beltzer
Siemens (Friedrik)

Great Britain
Aston Flint Glass Works
Bibby Sterilin
Block Mills, Portsmouth
Bramah Security Equipment
Robert Chance
Coalbrookdale Company
David Mellor
Denby Pottery
Dunnill & Co.
Elkington & Co
Ford Motor Company, Manchester
Francis Needham
Haws Eliot
Hawthorn, Newcastle
Humber Bicycle Co.
E. Hunter
Keep & Hinckley
Marples
Marriot & Cooper
Maw & Co.

Milton Brook Pottery
M.R. & Co.
J. Nasmyth
Rockware Glass
Smiths, Hurstmonceux
Starley
Swan
Taylor Law & Co.
T.G.Green Pottery
Walter Macfarlan & Co.
Wedgwood
Whitworth & Co.
Winfield & Co.
W. & G. Wynn

Italy
Vulcania

Japan
Fujii Metal

Norway
Nøstetangen Glasverk

Sweden
Årnäs
Höganäs
Mariebjerg Fajancefabrik

U.S.A
Edison
Ford Motor Co.
Waterbury Clock Co.

MUSEUMS
and industrial archaeological sites.

Recording work has been carried out with the co-operation of these institutions:

Denmark
Aalborg Historical Museum
Automobile Museum, Gjern
Clausholm Manor, nr. Hadsten
Danish Museum of Electricity, nr. Bjerringbro
Denmark's Museum of Technology, Elsinore
Louisiana Museum of Art, Humlebæk
Kunstindustrimuseet (The Museum of Decorative Art),
    Copenhagen
Museum of Cultural History, Randers
Museum of workers, craftsmen, and industry, Horsens
Old Town, Museum of Urban Culture, Aarhus
Ordrupgaard Museum, nr. Copenhagen
Royal Library, Copenhagen
Silkeborg Museum, Silkeborg
State Library, Aarhus
Sundby Collection, nr. Aalborg

France
Conservatoire National des Arts et Metiers, Paris

Germany
Deutchrs Museum, Munich
Die Neue Sammlung, Munich

Great Britain
Abbeydale Museum, Sheffield
Avery Museum of Weighing, Soho Foundry, Smethwick
Beamish Museum, Co. Durham
Birmingham City Museum
Black Country Museum, Dudley
Canal Museum, Stoke Bruerne
Chair Museum, High Wycombe
City Museum and Art Gallery, Hanley, Stoke-on-Trent
Forge Mill, Redditch
Gladstone Pottery Museum, Stoke-on-Trent
Ironbridge Gorge Museum Region, Telford
Killhope Lead Crushing Mill, Durham
Local Studies Department, Birmingham Library
London Transport Museum, Covent Garden
Museum of Science and Industry, Manchester
Naval Dockyards, Portsmouth
New Lanark, industrial community, Scotland
North Wales Quarrying Museum, Gwynedd
Quarry Bank Mill, Styal, Cheshire
Ryhope Pumping Station, Sunderland
Science Museum, London
Sheffield Museum
Southsea Castle Museum
Tide Mill, Woodbridge
Victoria and Albert Museum, London
Wedgwood Museum, nr. Stoke-on-Trent
Wortley Forge nr. Sheffield

Sweden
Höganäs Museum, Höganäs
Museum of Technology, Stockholm

BIBLIOGRAPHY
and further reading.

Books

Andersen, K.O., *Jern og Metalindustriens Materialer,* Teknisk Skoleforenings Forlag, Copenhagen 1961.

Andrews, Edward Deming, and Faith Andrews, *Shaker Furniture,* Dover Publications, New York 1950.

Anthony, John, *Joseph Paxton, 1803-1865,* Shire Publications, Princes Risborough 1992.

Austwick, J.&B., *The Decorated Tile,* Pitman House, London 1980.

Bartram, Alan, *Street Name Lettering in the British Isles,* Lund Humphries, London 1978.

Bennet, Arnold, books about *The Five Towns* (which make up Stoke-on-Trent), Penguin, Harmondsworth.

Bracegirdle, Brian, *The Archæology of the Industrial Revolution,* Heinemann, London 1974.

Bracegirdle, Brian and Patricia H. Miles, *Thomas Telford,* David & Charles, Newton Abbot 1973.

Bracegirdle, Brian and Patricia H. Miles, *The Darbys and the Iron Bridge,* David & Charles, Newton Abbot 1974.

Brown, John, *Welsh Stick Chairs,* Abercastle Publications, Newport 1990.

Bruton, Eric, *The History of Clocks and Watches,* Crescent Books, New York 1979.

Buchanan, R.A., *Industrial Archaeology in Britain,* Penguin Books, Harmondsworth 1974.

Burton, Anthony, *Remains of a Revolution,* André Deutsch, London 1975.

Burton, Anthony, *Our Industrial Past,* George Philip in association with The National Trust, London 1983.

Carrington, Noel and Clark Hutton, *Popular Art in Britain,* King Penguin series, Penguin, Harmondsworth 1945.

Christiansen, Povl and Hakon Stephensen, *The Craftsmen Show the Way* - Cabinetmakers Guild exhibitions 1927-1966, Copenhagen 1966.

Crowley, T.E., *Beam Engines,* Shire Publications, Princes Risborough 1976.

Derry, T.K. and Trevor Williams, *A Short History of Technology,* Oxford University Press, London 1973.

Diderot, Denis, *A Diderot Pictorial Encyclopedia of Trades and Industry,* edited by Charles Coulston Gillispie, 2 vols. Dover Publications, New York 1959.

Diderot, Denis, *Diderot Encyclopedia, the Complete Illustrations 1762-1777.* Editor: Arnoldo Mondadori. 4 vols. & index. Harry N. Abrams, New York 1978.

Ditzel, Nanna and Jørgen Ditzel, *Møbel Tegninger (trans. Furniture Drawings),* Frederiksberg Tekniske Skole, 1950. (knowledge of Danish not necessary).

Ditzel, Nanna and Jørgen Ditzel, *Danish Chairs,* Høst & Søns Forlag 1954.

Exner, Vilhelm Franz, *Das Biegen des Holzes,* Leipzig 1922. (a description in German of the Thonet wood bending technique).

Ford, Henry, *My Life and Work,* William Heinemann, London 1925.

Fournier, Robert, *Illustrated Dictionary of Practical Pottery,* A&C Black, London 1992.

Frederiksen, Erik Ellegaard, *Knud V. Engelhardt, Arkitekt & Bogtrykker, 1882-1931.* Arkitektens Forlag, Copenhagen 1965. (Danish, well-illustrated).

Frederiksen, Erik Ellegaard, Niels Kryger and Vibeke Lassen Nielsen, *Jens Nielsen,* Danish Design Center, Copenhagen, 1996.

George Allen & Unwin, *How Things Work,* vols.1&2, London 1977.

BIBLIOGRAPHY

Giedion, Siegfried, *Mechanization Takes Command*, W.W.Norton & Co., New York 1969.

Gloag, John, *Dictionary of Furniture*, Unwin Hyman, London 1990.

Goodman, W.L., *The History of Woodworking Tools*, G.Bell & Sons, London 1971.

Goodman, W.L., *British Plane Makers from 1700*, G. Bell & Sons, London 1968.

Griffith, Samuel, *Guide to the Iron Trade of Great Britain*, 1873. David & Charles, Newton Abbot 1967.

Grönstrand, Satu, *Alvar and Aino Aalto as glass designers*, Iittala-Nuutajärvi Oy 1988.

Gwyther, J.L. and R.V. Page, *An Introduction to Workshop Processes*, Penguin library of technology, Penguin Books, Harmondsworth 1968.

Hamilton, David, *Architectural Ceramics*, Thames & Hudson, London 1978.

Heskett, John, *Industrial Design*, Thames & Hudson, London 1980.

Hounshell, David A., *From the American System to Mass Production 1800 - 1932*, The Johns Hopkins University Press, Baltimore 1985.

Hunt, Robert, *Ure's Dictionary of Arts, Manufactures, and Mines*, 4 volumes, Longmans, Green, &Co. London 1878.

Jagger, Cedric, *The Worlds Great Clocks and Watches*, Hamlyn, London 1977.

Jalk, Grete, *40 Years of Danish Furniture Design, the Copenhagen Cabinetmakers' Guild Exhibitions 1927-1966*. Teknological Institute, Copenhagen 1987.

Johanssen, Ebbe, *Kakkelovn og jernovn* (trans. *Ceramic stove and iron stove)* Nyt Nordisk Forlag Arnold Busck, Copenhagen 1980. (Danish, well illustrated).

Johanssen, Ebbe, *Danske antikviteter af støbejern* (trans. *Danish antiques of cast iron)* Nyt Nordisk Forlag Arnold Busck, Copenhagen 1982. (Danish, well illustrated).

Jones, Edward, and Christopher Woodward, *A Guide to the Architecture of London*, Weidenfeld & Nicolson, London 1983.

Karlsen, Arne, *Furniture Designed by Børge Mogensen*, The Danish Architectural Press, Copenhagen 1968. *Dansk Møbel Kunst i det 20. århundrede* (trans. *Danish Furniture Art in the 20th century)*, Christian Ejlers, Copenhagen, vol. 1 1990, vol. 2 1991. (Danish, well illustrated).

Karlsen, Arne and Anker Tiedemann, *Made in Denmark*, Jul Gjellerups Forlag, Copenhagen 1960.

Lassen, Erik, *Ske, Kniv og Gaffel / Spoon, Knife And Fork*, Høst &Søn, Fredericia 1960. (In Danish and English).

Law, R.J., *The Steam Engine*, Her Majesty's Stationery Office, London 1965. (booklet).

Logie, Gordon, *Furniture from Machines*, George Allen and Unwin, London 1947.

Lucie-Smith, Edward, *A History of Industrial Design*, Phaidon, Oxford 1983.

Lütken, André and Helge Holst, *Opfindelsernes Bog*, and Holst, Helge, *Opfindernes Liv*, 5 vols, Nordisk Forlag, Copenhagen 1912.

MacCarthy, Fiona, *A History of British Design 1830-1970*, George Allen & Unwin, London 1979.

McNeil, Ian, *Joseph Bramah, a century of invention 1749 – 1851*, David & Charles, Newton Abbot 1968.

Mellor, David, *Kitchen Guide - A good cook's catalogue*, London.

Mumford, Lewis, *Technics and Civilization*, George Routledge, London 1934.

Palmer, Brooks, *The Book of American Clocks*, The Macmillan Company, New York 1950.

Palmer, Roy, *The Water Closet*, David & Charles, Newton Abbot 1973.

Petersen, Bernt, with Poul Hansen, Einer Pedersen, Kristian Jacobsen and Marianne Wegner Sørensen, *Hans J. Wegner, en stolemager* (trans. *Hans J. Wegner, a chairmaker)*. Danish Design Center 1989. 75th anniversary exhibition of Wegner's work in book form. (Knowledge of Danish not necessary).

Pye, David, *The Nature of Design*, Studio Vista, London 1972.

Pye, David, *The Nature and Art of Workmanship*, A & C Black, London 1995.

Rasmussen, Steen Eiler, *Britisk Brugskunst*, (trans. *British Product Design)*. A selection of pictures from the 1932 exhibition at the Museum of Applied Art in Copenhagen, 1933. (Rasmussen's evaluations of the objects are in Danish, but the photographs are good).

Read, Herbert, *Art and Industry*, Faber & Faber, London 1944.

Reilly, Robin, *Wedgwood*, vols.1 & 2, Macmillan, London 1989.

Rees, A., *The Cyclopaedia of Arts, Sciences, and literature*, London 1819.

Richards, J.M., *The Functional Tradition*, The Architectural Press, London, 1968.

Richards, J.M., *Who's Who in Architecture, from 1400 to the present day*, Weidenfeld and Nicolson, London 1977.

Roe, F.Gordon, *Windsor Chairs*, Phoenix House, London 1953.

Rolt, L.T.C., *Tools for the job*, Batsford, London 1965. *Isambard Kingdom Brunel*, 1972. *Victorian Engineering*, 1974, both Penguin, Harmondsworth.

Rowland, K.T., *Eighteenth Century Inventions*, David & Charles, Newton Abbot 1974.

Salaman, R.A., *Dictionary of Tools used in the Woodworking Trades 1700-1970*, George Allen & Unwin, London 1975.

Sano, Yuji, *The Scissors Book*, Japan.

Schaefer, Herwin, *The Roots of Modern Design, functional tradition in the 19th century*, Studio Vista, London 1970.

Schlüter, Mogens, *Danske flasker*, (trans. *Danish bottles)* From the renaissance to our day, Nyt Nordisk Forlag Arnold Busck, Copenhagen 1984. (Danish, well illustrated).

Siek, Frederik, *Contemporary Danish Furniture Design*, Nyt Nordisk Forlag Arnold Busck, Copenhagen 1990.

Sloane, Eric, *A Museum of Early American Tools*, Ballantine Books, New York 1973 .

Smiles, Samuel, *The Lives of the Engineers* (2 vols.), Murray, London 1861.

Smith, Stuart, *A View from the Iron Bridge*, Iron Bridge Gorge Museum Trust, 1979.

Sparkes, Ivan, *The Windsor Chair*, Spurbooks, Bourn End 1975.

Sprigg, June, *Shaker Design*, Whitney Museum of American Art, in association with W.W.Norton, New York 1986.

Sprigg, June and David Larkin, *Shaker Life, Work, and Art*, Cassell, London 1994.

Steeds, W., *A History of Machine Tools 1700-1910*, Oxford University Press, London 1969.

Strandh, Sigvard, *A History of the Machine*, A&W Publishers, New York 1979. (AB Nordbok, Sweden).

Street, Arthur and William Alexander, *Metals in the Service of Man*, Penguin, Harmondsworth, 1989.

Swedberg, Robert W. and Harriet Swedberg, *American Clocks and Clockmakers*, Radnor 1989.

Thau, Carsten and Kjeld Vindum, Arne Jacobsen, Arkitektens Forlag, Copenhagen 1998

Timmins, Samuel, *The Resources, Products and Industrial History of Birmingham and the Midland Hardware District*, Robert Hardwicke, London 1866.

Tomlinson, Charles and A.C.Hobbs, *The Construction of Locks*, Virtue and Co., London 1868.

Vegesack, Alexander von, *Das Thonet Buch*, Bangert Verlag, München 1987.

Wanscher, Ole, *Møbelkunsten*, Thaning & Appels Forlag, Copenhagen 1955. *The Art of Furniture*, Allen & Unwin, London 1968.

Wilkes, Lyall, *John Dobson, Architect and Landscape Gardener,* Oriel Press, Stocksfield 1980.

Windsor, Alan, *Peter Behrens,* London 1981.

Wright, Lawrence, *Clean and Decent,* London 1960.

Research and compendia

Architectural Students Association Journal, *Plan No.5,* London 1949.

Danmarks Designskole, *Opmålingsstudier, Kunstindustrmuseets Stolesamling* (trans. *Measured Drawings of Chairs at the Museum of Applied Arts),* Copenhagen 1991.

Aarhus School of Architecture, Department of furniture, interior, and industrial design: *Göta Kanal – Svensk funktionsbestemt arkitektur i 1800tallet.* 1976. (Danish, well illustrated).

Gudiksen, Marno, *The Cricket Bat,* Arkitekten (periodical) nr.13 September 1995. (Danish, well illustrated. Analysis of its construction and materials).

Guthrie, J.L., A. Allen and C.R.Jones, 'Restoration of the Palm House at the Royal Botanic Gardens, Kew'. Article in *Proceedings of the Institute of Civil Engineers,* Part 1, Design and Construction 1988.

Heath, Adrian, and Aage Lund Jensen, *Early Industrial Design in England,* Aarhus School of Architecture, 1974.

Heath, Adrian, *Noter om Metal materialer,* (trans. *Notes on Metals)* Aarhus School of Architecture, 1980.

Heath, Dittc, *Noter om Træ,* (trans. *Notes on Wood*), Aarhus School of Architecture, 1983.

Museum & exhibition literature

Atkinson, Frank, *Industrial Archaeology, Top Ten Sites in North East England,* Frank Graham, Newcastle 1971.

Die Neue Sammlung publications, Munich:
Beutler, Christian, *Weltausstellungen im 19. Jahrhundert,* 1973.
Fischer, Wend, *Die Verborgene Vernuft.*
Fischer, Wend, *Die Neue Sammlung.*
Hartmann, G.B.von and Wend Fischer, *Zwischen Kunst und Industrie der Deutche Werkbund,* 1975.
Mang, Karl and Wend Fischer, *The Shakers,* 1974.

Finlands Arkitekt Museum, Konstflitföreningen i Finland, Museet för nutidskonst (3 organisations), *Sagt i Trä* (trans. *Said in Wood),* 228 page exhibition catalogue of large photos, 1987.

Heath, Adrian, Ditte Heath and Aage Lund Jensen, *The Historical Development of Industrial Design,* catalogue of an exhibition of photographs at the Royal College of Art, London 1982.

Heath, Adrian, *Furniture as a Cultural Factor,* article in the catalogue of Maker Designers Today exhibition at Camden Arts Centre, 1984.

Iron Bridge Gorge Museum, published 1973-1979:
1. The Iron Bridge
2. The Coalbrookdale Museum
3. Coalport, New Town of the 1790s
4. Coalport China Works Museum
5. Blists Hill Open Air Museum
7. The Tar Tunnel

Official descriptive and illustrated Catalogue of the *Great Exhibition 1851,* Vols. I, II, III, London 1851.

Ryhope Pumping Station's leaflet of 1974.

The Institution of Civil Engineers, *Thomas Telford,* the bicentenary exhibition booklet, 1957.

Articles

Heath, Adrian, Ditte Heath and Aage Lund Jensen, *Den Historiske Udvikling af Industriel Design* (trans. The Historical Development of Industrial Design), Arkitekten Nr.11, June 1982. Well illustrated.

Petersen, Gunnar Bilman, *Kakkelovne* (trans. *Heating stoves).* Well illustrated article in Danish in the periodical Arkitekten, Copenhagen 1923.

Firm's literature

F.D.B. (Danish Cooperative Society), Furniture catalogues from 1950 onwards.

Fritz Hansen, furniture catalogues from 1951 onwards.

Jepsen, Anton, *Danske snedkermøbler gennem 125 år (*Trans. *Danske Cabinetmaker Furniture for 125 years).* Short history of the firm of Rudolph Rasmussen, and a selection of furniture photographed in the firm's permanent exhibition in Copenhagen, 1994.

Møller, Viggo Sten, and Johannes Hansen, *Vore Dages Møbler. Et Håndværks Vej i en Maskintid.* (Trans. *Furniture Today. The Way of a Craft in a Machine Age.* Published by Master Cabinetmaker Johannes Hansen for the workshop's 40th anniversary, *c.*1953.

Thonet Bentwood & other Furniture, the 1904 illustrated catalogue. Introduction by Christopher Wilk. Dover Publications, New York 1980.

INDEX